Lecture Notes in Mechanical Engineering

Lecture Notes in Mechanical Engineering (LNME) publishes the latest developments in Mechanical Engineering—quickly, informally and with high quality. Original research or contributions reported in proceedings and post-proceedings represents the core of LNME. Volumes published in LNME embrace all aspects, subfields and new challenges of mechanical engineering.

To submit a proposal or request further information, please contact the Springer Editor of your location:

Europe, USA, Africa: Leontina Di Cecco at Leontina.dicecco@springer.com
China: Ella Zhang at ella.zhang@cn.springernature.com
India, Rest of Asia, Australia, New Zealand: Swati Meherishi at swati.meherishi@springer.com

Topics in the series include:

- Engineering Design
- Machinery and Machine Elements
- Mechanical Structures and Stress Analysis
- Automotive Engineering
- Engine Technology
- Aerospace Technology and Astronautics
- Nanotechnology and Microengineering
- Control, Robotics, Mechatronics
- MEMS
- Theoretical and Applied Mechanics
- Dynamical Systems, Control
- Fluid Mechanics
- Engineering Thermodynamics, Heat and Mass Transfer
- Manufacturing Engineering and Smart Manufacturing
- Precision Engineering, Instrumentation, Measurement
- Materials Engineering
- Tribology and Surface Technology

Indexed by SCOPUS, EI Compendex, and INSPEC.

All books published in the series are evaluated by Web of Science for the Conference Proceedings Citation Index (CPCI).

To submit a proposal for a monograph, please check our Springer Tracts in Mechanical Engineering at https://link.springer.com/bookseries/11693.

Francisco J. G. Silva · Vitalii Ivanov ·
Arnaldo Manuel Guedes Pinto ·
Rita de Cassia M. Sales-Contini
Editors

Advances in Design, Simulation and Manufacturing VIII

Proceedings of the 8th International Conference on Design, Simulation, Manufacturing: The Innovation Exchange, DSMIE-2025, June 17–20, 2025, Porto, Portugal - Volume 3: Materials Engineering and Engineering Education

Springer

Editors
Francisco J. G. Silva
School of Engineering
Polytechnic Institute of Porto
Porto, Portugal

Arnaldo Manuel Guedes Pinto
Polytechnic Institute of Porto
Porto, Portugal

Vitalii Ivanov
Department of Manufacturing Engineering, Machines and Tools
Sumy State University
Sumy, Ukraine

Rita de Cassia M. Sales-Contini
Polytechnic Institute of Porto
Porto, Portugal

ISSN 2195-4356 ISSN 2195-4364 (electronic)
Lecture Notes in Mechanical Engineering
ISBN 978-3-031-96412-1 ISBN 978-3-031-96413-8 (eBook)
https://doi.org/10.1007/978-3-031-96413-8

This Springer imprint is published by the registered company Springer Nature Switzerland AG
The registered company address is: Gewerbestrasse 11, 6330 Cham, Switzerland

Preface

This volume of Lecture Notes in Mechanical Engineering contains selected papers presented at the 8th International Conference on Design, Simulation, Manufacturing: The Innovation Exchange (DSMIE-2025), held in Porto, Portugal, on June 17–20, 2025. The conference was organized by the International Association for Technological Development and Innovations, Polytechnic of Porto, and Sumy State University, and in partnership with Poznan University of Technology (Poland), Technical University of Kosice (Slovak Republic), University of West Bohemia (Czech Republic), Kielce University of Technology (Poland), Association for Promoting Innovative Technologies–Innovative FET (Croatia), and Society for Robotics of Bosnia and Herzegovina (Bosnia and Herzegovina).

DSMIE Conference Series is the international forum for fundamental and applied research and industrial applications in engineering. The conference focuses on research challenges in the fields of Manufacturing Engineering, Materials Engineering, Mechanical Engineering, and Industrial Management, addressing current and future trends in design approaches, simulation techniques, and manufacturing technologies, highlighting the growing role of smart manufacturing systems, artificial intelligence, standards-based integration, and innovations implementation in the transition to sustainable, human-centric and resilient engineering solutions. DSMIE brings together researchers from academic institutions, leading industrial companies, and government laboratories worldwide to promote and popularize the scientific fundamentals of engineering.

DSMIE-2025 received 227 contributions. After a two-stage single-blind review, the Program Committee accepted 86 papers written by 362 authors from 31 countries. Thank you very much to the authors for their contribution. Each paper was peer-reviewed by at least three members of the Program Committee, along with invited external reviewers. Each submission was evaluated for originality, scientific significance, research methodology, accuracy of the research content, discussion of the obtained results, validity of the conclusions, completeness, clarity of presentation, and overall readability. The originality of the submissions was checked by Turnitin plagiarism software. The selected papers are published in the present book, achieving an acceptance rate of about 38%.

We would like to thank members of the Program Committee and invited external reviewers for their efforts and expertise in contributing to reviewing, without which it would be impossible to maintain the high standards of peer-reviewed papers. 115 Program Committee members and 14 invited external reviewers devoted their time and energy to peer reviewing manuscripts. Our reviewers come from around the world, representing 22 countries, and are affiliated with 39 institutions.

Extended versions of selected best papers will be published in scientific journals: International Journal of Advanced Manufacturing Technology (Springer), Mechanics Based Design of Structures and Machines (Taylor and Francis), Production & Manufacturing Research (Taylor and Francis), Journal of Engineering Manufacture (SAGE),

Journal of Manufacturing and Materials Processing (MDPI), and Machine Learning and Knowledge Extraction (MDPI).

Thank you very much to the keynote speakers: Prof. Luis Fonseca (Polytechnic of Porto, Portugal), Prof. Margherita Peruzzini (University of Bologna, Italy), Prof. Paul-Eric Dossou (Institut Catholique d'Arts et Metiers-ICAM, France), Prof. Panorios Benardos (National Technical University of Athens, Greece), and Grzegorz Krolczyk (Opole University of Technology, Poland).

We sincerely appreciate the invited speaker of the DSMIE Workshop "Agile in Manufacturing Processes": Dr. Bohdan Haidabrus (Accenture Baltics; Riga Technical University, Latvia).

The book "Advances in Design, Simulation and Manufacturing VIII" was organized in four LNME volumes according to the main conference topics: Volume 1—Manufacturing Engineering, Volume 2—Mechanical Engineering, Volume 3—Materials Engineering, and Volume 4—Industrial Management. Each volume is devoted to research studies in design, simulation, and manufacturing within the main conference areas.

We appreciate the partnership with Springer Nature, Turnitin, ConfTool, and our sponsors for their essential support during the preparation of DSMIE-2025.

Thank you very much to the DSMIE Team. Their outstanding involvement and hard work were crucial to the success of the conference.

DSMIE's motto is *"Together we can do more for science, technology, engineering, and education"*.

June 2025

Francisco J. G. Silva
Vitalii Ivanov
Arnaldo Manuel Guedes Pinto
Rita de Cassia M. Sales-Contini

Organization

Steering Committee

General Chair

Vitalii Ivanov	Sumy State University, Ukraine

Co-chair

Francisco J. G. Silva	Polytechnic of Porto, Portugal

Co-chairs, Program Committee and Publication Chairs

Raul Duarte Campilho	Polytechnic of Porto, Portugal
Oleksandr Liaposhchenko	Sumy State University, Ukraine
Justyna Trojanowska	Poznan University of Technology, Poland

Co-chairs, Organizational Chairs

Oleksandr Gusak	Sumy State University, Ukraine
Khrystyna Berladir	Sumy State University, Ukraine
Teresa Pereira	Polytechnic of Porto, Portugal
Arnaldo Guedes Pinto	Polytechnic of Porto, Portugal

Program Committee

Katarzyna Antosz	Rzeszow University of Technology, Poland
Genevieve Bakam	Tshwane University of Technology, South Africa
Liliana-Laura Badita-Voicu	National Institute of Research and Development in Mechatronics and Measurement Technique, Romania
Andresa Baptista	Polytechnic of Porto, Portugal
Marta Barbosa	University of Porto, Portugal
Iryna Bashynska	AGH University of Science and Technology, Poland

Yevheniia Basova	National Technical University "Kharkiv Polytechnic Institute", Ukraine
Andre Batako	Liverpool John Moores University, UK
Seyedasghar Bayatghiasi	University of Aveiro, Portugal
Annamaria Behunova	Technical University of Kosice, Slovak Republic
Khrystyna Berladir	Sumy State University, Ukraine
Pietro Bilancia	University of Modena and Reggio Emilia, Italy
Ricardo Branco	University of Coimbra, Portugal
Andrii Breus	National Aerospace University–Kharkiv Aviation Institute, Ukraine
Anna Burduk	Wroclaw University of Science and Technology, Poland
Suthep Butdee	King Mongkut's University of Technology, Thailand
Michele Cali	University of Catania, Italy
Emilia Campean	Technical University of Cluj-Napoca, Romania
Raul Duarte Campilho	Polytechnic of Porto, Portugal
Vasile George Cioata	Polytechnic University of Timisoara, Romania
Olaf Ciszak	Poznan University of Technology, Poland
Ruben Costa	Polytechnic of Porto, Portugal
Radu Cotetiu	Technical University of Cluj-Napoca, Romania
Rita Sales-Contini	Polytechnic of Porto, Portugal
Predrag Dasic	High Technical Mechanical School of Professional Studies, Trstenik, Serbia
Ivan Dehtiarov	Sumy State University, Ukraine
Yuliia Denysenko	Sumy State University, Ukraine
Oleksandr Derevianchenko	Odesa Polytechnic National University, Ukraine
Sergiy Dobrotvorskyy	University of Life Science, Poland
Tygran Dzuguryan	Maritime University of Szczecin, Poland
Milan Edl	University of West Bohemia, Czech Republic
Luis Ferreira	Polytechnic of Porto, Portugal
Isabel Perdigão	Figueiredo Polytechnic of Porto, Portugal
Mykhaylo Frolov	National University "Zaporizhzhia Polytechnic", Ukraine
Alexandra Gavina	Polytechnic of Porto, Portugal
Renata Gnatowska	Czestochowa University of Technology, Poland
Arkadiusz Gola	Lublin University of Technology, Poland
Janis Grabis	Riga Technical University, Latvia
Jakub Krzysztof Grabski	Poznan University of Technology, PolandLublin University of Technology, Poland
Marek Gucwa	Czestochowa University of Technology, Poland
Volodymyr Gurey	Lviv Polytechnic National University, Ukraine
Bohdan Haidabrus	Riga Technical University, Latvia

Nataliia Yevtushenko	National Technical University "Kharkiv Polytechnic Institute", Ukraine
Oleg Zabolotnyi	Lutsk National Technical University, Ukraine
Nataliia Zaichuk	Lutsk National Technical University, Ukraine
Volodymyr Zavialov	National University of Food Technologies, Ukraine
Przemyslaw Zawadzki	Poznan University of Technology, Poland

Invited External Reviewers

Orhan Cakir	Yildiz Technical University, Turkey
Oguz Colak	Eskisehir Technical University, Turkey
Roland Iosif Moraru	University of Petrosani, Romania
Sorin Aurel Ratiu	Polytechnic University of Timisoara, Romania
Joanna Rosak-Szyrocka	Czestochowa University of Technology, Poland
Lesya Shkitsa	Ivano-Frankivsk National Technical University of Oil and Gas, Ukraine
Natalya Shramenko	Lviv Polytechnic National University, Ukraine
Robert Sika	Poznan University of Technology, Poland
Dusan Simsik	Technical University of Kosice, Slovak Republic
Yuliia Tarasevych	AGH University of Science and Technology, Poland
Ovidiu Tirian	Polytechnic University of Timisoara, Romania
Anatolii Tkachuk	Lutsk National Technical University, Ukraine
Nicolae Ungureanu	Technical University of Cluj-Napoca, Romania
Justyna Zywiolek	Czestochowa University of Technology, Poland

DSMIE Team

Andresa Baptista	Polytechnic of Porto, Portugal
Marta Barbosa	Polytechnic of Porto, Portugal
Khrystyna Berladir	Sumy State University, Ukraine
Raul Duarte Campilho	Polytechnic of Porto, Portugal
Luis Filipe Coelho	Polytechnic of Porto, Portugal
Ruben Costa	Polytechnic of Porto, Portugal
Yuliia Denysenko	Sumy State University, Ukraine
Luis Pinto Ferreira	Polytechnic of Porto, Portugal
Alexandra Gavina	Polytechnic of Porto, Portugal
Oleksandr Gusak	Sumy State University, Ukraine
Vitalii Ivanov	Sumy State University, Ukraine

Contents

Protective Technologies

Antibacterial Properties of the Biodegradable 3D-Printed PCL/PLA Scaffolds with Silver Nanoparticles

Oleksandr Oleshko[1], Khrystyna Berladir[1,2](✉), Tetiana Oleshko[1], Oleksandr Korol[1], and Răzvan Păcurar[3]

[1] Sumy State University, 116, Kharkivska Street, Sumy 40007, Ukraine
khrys.berladir@gmail.com
[2] Technical University of Kosice, 1, Bayerova Street, 08001 Presov, Slovak Republic
[3] Technical University of Cluj-Napoca, B-Dul Muncii 103-105, 400641 Cluj-Napoca, Romania

Abstract. This study investigated the antibacterial properties of biodegradable 3D-printed scaffolds composed of polylactic acid (PLA) and polycaprolactone (PCL) with silver nanoparticles (AgNPs). The scaffolds were fabricated using fused filament fabrication 3D printing, and their antibacterial efficacy was tested against *Staphylococcus aureus* (strain B 918). The experimental samples included varying ratios of PLA and PCL, with and without AgNPs. The bacterial adhesion and proliferation were assessed at 4, 6, 24, and 48 h, revealing that scaffolds with AgNPs exhibited significantly reduced bacterial growth compared to controls. At 24 h, bacterial density on PCL30-PLA70-AgNPs was 5.70 ± 0.13 log CFU, significantly lower than the 8.00 ± 0.14 log CFU observed on PCL30-PLA70. Furthermore, degradation tests in simulated body fluid confirmed a correlation between mass loss and bacterial inhibition, suggesting a controlled ion release mechanism. The results demonstrate that the incorporation of AgNPs in PLA/PCL scaffolds creates their antibacterial properties while maintaining biodegradability, making them promising candidates for bone tissue engineering and dental implants.

Keywords: Additive Manufacturing · Tissue Engineering · Biomaterials · Polymers · Bacterial Adhesion · Process Innovation

1 Introduction

Research into biodegradable 3D-printed scaffolds is an important direction in modern engineering as they combine innovative technologies with environmental and medical needs, making this field interdisciplinary. The current state of research in the field of biodegradable nanocomposite materials [1-3] indicates their great potential in tissue engineering, as they can mimic natural tissue structures. A promising direction that contributes to a breakthrough in the field of bioengineering is the creation of polymer biodegradable nanocomposites through direct 3D printing, which combines innovative technologies of materials science and additive manufacturing [4]. Design and computer modeling technologies allow the adjustment of the structural features and functional properties of composites by controlling printing parameters [5].

F. J. G. Silva et al. (Eds.): DSMIE 2025, LNME, pp. 3–14, 2025.
https://doi.org/10.1007/978-3-031-96413-8_1

This work is devoted to studying the antibacterial properties of functionalized surfaces of designed biodegradable scaffolds based on a mixture of polylactic acid (PLA) and polycaprolactone (PCL) polymers with the addition of silver nanoparticles manufactured using additive 3D printing technologies.

2 Literature Review

It is known that silver nanoparticles (AgNPs) are characterized by a pronounced bactericidal effect on eukaryotic cells. This plays an important role in the osseointegration processes. Therefore, silver nanoparticles have proven to be widely used in medicine for restoring bone structure [6] and dental implants [7]. The widespread use of silver nanoparticles in medicine, orthopaedics, and dentistry is due to their antibacterial properties targeting a wide range of gram-positive and gram- negative bacteria [8]. However, Zhang et al. [9] have found that high concentrations of AgNPs have the ability to inhibit osseointegration and have a cytotoxic effect. This leads to the conclusion that there is a need for a balance in the concentration of silver nanoparticles. Myakinin et al. [10], from the experiment with osteoblast cell culture, demonstrated that the surface with silver nanoparticles is safe and non-toxic. Prevention of the development of cytotoxic effects is possible in the case of selecting balanced concentrations of AgNPs, ensuring a low rate of release of silver ions during long-term use of the implant. A high level of antibacterial activity against gram-positive bacteria was also found. Thus, silver nanoparticles do not significantly affect the growth rate of osteoblast cells and their metabolic activity, as assessed by the method of determining collagen formation.

The mechanism of antibacterial action of nanoparticles has not yet been fully understood, so this process is ongoing. There is data confirming the role of metal nanoparticles in the formation of reactive oxygen species and the development of oxidative stress in bacterial cells, which leads to membrane destruction and death. Besinis et al. [11] analyzed the antibacterial activity of nanoparticles of various metals used for dental implantation. As a result of the comparison, it was found that the intensity of the antibacterial action of Ag + nanoparticles was higher than that of all other studied samples.

The research groups [12-14] concluded that materials used for bone tissue engineering and dental implants must be biocompatible with cells of the human body, must have the ability for bioactivity and osteoconduction, and must be biodegradable. The presence of such disadvantages as the high level of rigidity of structures made of metal, the ability of metals to enter chemical reactions with enzymes and release ions from the metal surface, toxicity, susceptibility to rapid corrosion, and the accession of concomitant infection makes metals unpopular when choosing a material for bone engineering and dental implants as noted in [15].

The need for materials with better properties has drawn the attention of the scientific community to polymer materials [16]. Among them, PLA and PCL have become popular because, in addition to their inherent number of advantages, synthetic polymers are not prone to denaturation during material processing and immunogenicity, like natural polymers, as described in [17, 18]. PCL and PLA are environmentally friendly biodegradable polymers with a high degree of biocompatibility, low melting point and viscosity, no toxic effects, and have the ability to regulate viscosity. They also have inherent mechanical and rheological characteristics, high crystallinity, and elasticity coefficient, which together contribute to the wide application of these biomaterials [19, 20].

Adding Ag + nanoparticles to PCL/PLA will allow obtaining material for repairing bone tissue damage or dental implants, which will have antibacterial properties immediately after installation in the defect site and throughout the degradation process. Shebi et al. [21] and Augustine et al. [22] have proved the antibacterial activity of PCL and PLA with the addition of silver against *E. Coli* and *S. aureus*. Thus, it is likely that PCL/PLA/AgNPs may be the material that will satisfy all the requirements for medical osteoreparative and dental implants. This substantiates the scientific novelty and value of this study: to design an innovative biodegradable PCL/PLA scaffold with silver nanoparticles with proven antibacterial properties.

3 Research Methodology

3.1 Synthesis of Materials

In this study, PLA (Fusion PLA 3d printing filament, signal white, 1.75 mm, standard TRD3D0020-1EA) and PCL (standard PCL50K-50MG) were used as raw materials for the manufacture of scaffolds. Silver nitrate ($AgNO_3$, 98%), poly(N-vinylpyrrolidone) (PVP, M_w = 40.000, K-25) were used for the synthesis of silver nanoparticles. Silicon carbide abrasive paper with different grain sizes was also used. All chemical reagents were purchased from Sigma-Aldrich (Darmstadt, Germany). Bacteriological media were purchased from HiMedia (Maharashtra, India). *Staphylococcus aureus B 918* were obtained from the National Collection of Microorganisms (D.K. Zabolotny Institute of Microbiology and Virology, Kyiv, Ukraine).

The method of material synthesis for both the control and experimental series is shown in Fig. 1. Silver nanoparticles were introduced into the structure of the material based on biodegradable polymers. The method of nanoparticle synthesis, namely its final stages, involves the use of high-temperature muffle furnaces, so the change in the characteristics of the latter during the extrusion process or 3D printing itself at a maximum temperature of 200 °C is eliminated. A detailed description of the method for synthesizing silver nanoparticles is described in [23].

3.2 3D Printing of Scaffolds

The marking and composition of the group of samples, both control (pure PLA and PCL) and experimental, are given in Table 1. The cylindrically shaped scaffolds (diameter 5 mm and height 5 mm) were created by 3D printing. The 3D model was developed in the FreeCad 0.19 program. After exporting the model in stl format, the finished model was checked for defects during modeling and printing using the Ultimaker Cura 4.11.0 software.

The printing parameters of the control samples are given in Table 2, and the experimental ones in Table 3. Support elements were not used. The obtained G-code for the Flyingbear Ghost 5 3D printer based on the model and printing parameters was transferred using a SanDisk SD card. Through the printer menu system, the selected file was

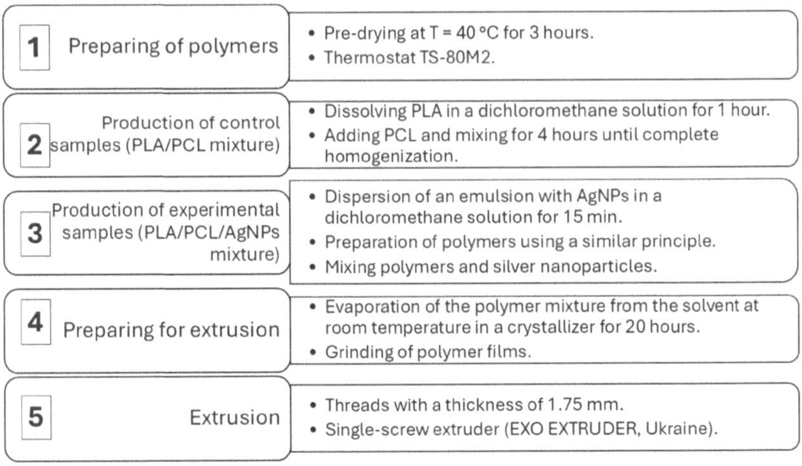

Fig. 1. Step-by-step synthesis technology of the tested materials.

Table 1. Marking and composition of control and experimental samples.

Marking	PCL (wt. %)	PLA (wt. %)
monopolymer PLA	–	100
monopolymer PCL	100	–
PCL 30–PLA 70	30	70
PCL 30–PLA 70 AgNPs	≈30	≈70
PCL 20–PLA 80	20	80
PCL 10–PLA 90	10	90

printed using the fused filament fabrication 3D printer FLYINGBEAR GHOST 5 in a closed room at an air temperature of 22 °C and a humidity of 45%. The printed samples were detached from the printer table after cooling with a spatula. To obtain statistically significant results, all studies were conducted with a mandatory 6-fold repetition. Data were presented as mean ± standard deviation.

3.3 Assessment of Bacterial Adhesion

To assess the antibacterial properties of the experimental material, the method of comparative analysis of the adhesion of microorganisms was used using the example of the gram-positive bacterium *Staphylococcus aureus* (strain B 918).

The bacterial strain was grown on nutrient agar for 24 h at 37 °C. This strain was initially suspended in saline (0.9% NaCl) and then resuspended in nutrient broth using McFarland standards until a density of 1×10^8 colony-forming units ml^{-1} (8 log CFU) was achieved. Cylindrical samples were incubated under static conditions for 4, 6, 24, and 48 h in a 24-well plate at 37 °C with 3.0 ml of bacterial suspension. After incubation,

Table 2. Printing parameters for control samples.

Material		PLA	PCL
Quality	Layer height	0.1 mm	0.1 mm
	The height of the first layer	0.1 mm	0.1 mm
Filling	Filling	100%	100%
	Template	Mesh	Mesh
Material	Nozzle temperature	205 (°C)	63 (°C)
	Table temperature	60 (°C)	30 (°C)
Speed	Printing	40 mm/s	15 mm/s
	Moving	120 mm/s	120 mm/s
	The first layer	15 mm/s	15 mm/s
Moving	The amount of rollback	6.5 mm	6.5 mm
	Retraction speed	35 mm/s	35 mm/s
Cooling		100%	100%
Type of adhesion		Classic	Classic

Table 3. Printing parameters for experimental samples.

Material	Print speed (mm/s)	Nozzle temperature (°C)	Table temperature (°C)
PCL 10 – PLA 90	25	200	60
PCL 20 – PLA 80	25	200	60
PCL 30 – PLA 70	50	200	60
PCL 30 – PLA 70 AgNPs	50	200	60

the samples with AgNPs were removed using sterile tweezers and were washed three times with 2.0 ml of sterile saline to remove non-adherent bacterial cells. The next step was to remove adhered bacteria from the functionalized surface of the test samples using an ultrasonic bath after immersion in sterile saline (1.0 ml) in a separate sterile tube. The specific number of CFU was counted using the strip plate technique after culturing 10 μl aliquots of saline from the sonicated tubes on a nutrient medium for 24 h. Scaffolds with combined copolymers were used as controls.

3.4 Test for Degradation

To assess the degradation in vitro of the control and experimental samples [24], a test was performed for immersion in a Simulated body fluid (SBF) solution, which, in terms of ion concentration, is as similar as possible to human blood plasma (Fig. 2).

Before the start of the study, the weight of all samples was determined using analytical-grade scales (Laboratory Electronic Balance AS 220.R2).

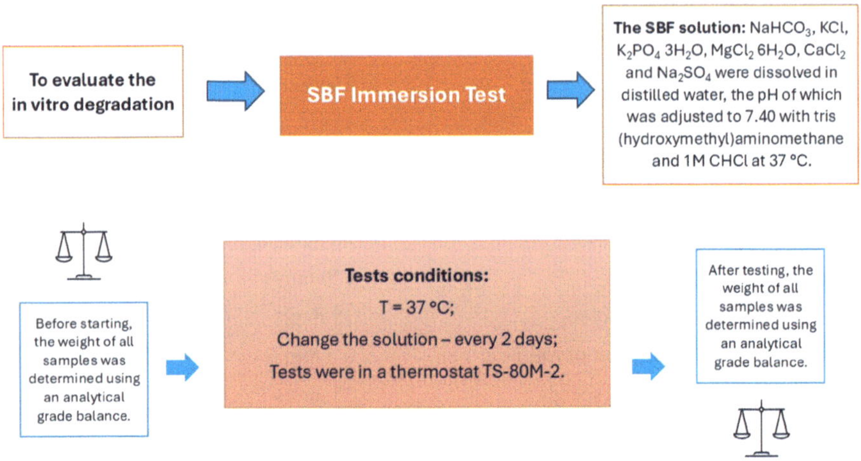

Fig. 2. Realization of the test for degradation.

A blood circulation simulation system was developed and patented (Fig. 3) for conducting the test under dynamic conditions.

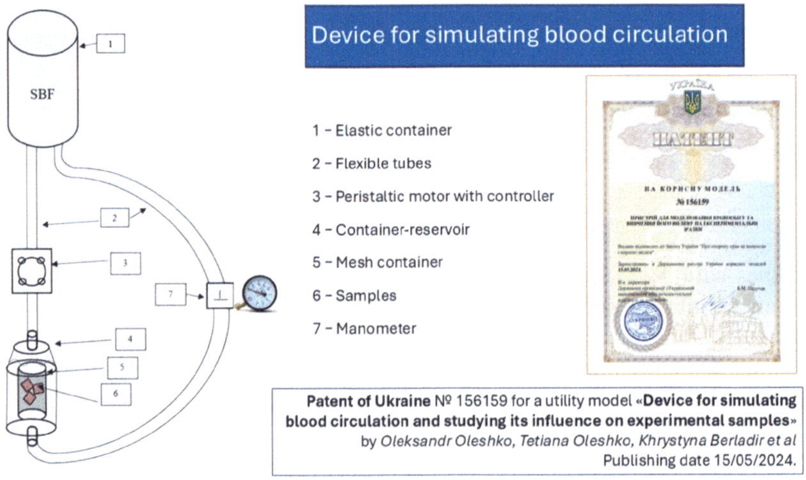

Fig. 3. Schematic representation of the circulatory simulation system.

SBF solution in a volume of 1 L was poured into an elastic container, and then through a system of medical silicone tubes, the liquid moved through a peristaltic motor with a speed controller to the container with samples and returned to the elastic container again. After closing the elastic container and starting the peristaltic motor according to the specified parameters, it obtained a filled artificially simulated circulatory system, which was hermetically sealed. Using a mechanical clamp, it compressed the elastic container, thereby, under the control of a manometer, increasing the pressure in the simulated circulatory system.

To ensure the stability of temperature parameters, the container with a circulatory simulation system was placed in a thermostat (Electric Dry Air Thermostat TC-80M-2) at a temperature of 37 °C, and the ventilation function was activated. A wide range of immersion times was selected, including 6, 24, 48, and 72 h and 7, 14, 21, and 28 days.

At the end of each period, the samples were washed with distilled water and dried on filter paper in a thermostat at 37 °C to constant weight (2 h), and a control weight measurement was performed. Weight loss was calculated using the formula:

$$\omega(\%) = (m_0 - m_1/m_0) \times 100$$

where m_0 – weight before immersion; m_1 – weight after immersion.

4 Results and Discussion

The results of the microbiological study confirmed the adhesion and proliferation of S. Aureus on the surface of both control and experimental samples after 4 h of cultivation and throughout the entire observation period of the experiment (Table 4).

Table 4. Assessment of the adhesion of microorganisms.

Material	Duration of cultivation, h			
	4	6	24	48
	Density, 1×10^8 CFU			
PCL30-PLA70	6.20 ± 0.15	6.90 ± 0.18	8.00 ± 0.14	8.00 ± 0.16
PCL30-PLA70-AgNPs	4.12 ± 0.09	5.10 ± 0.12	5.70 ± 0.13	5.23 ± 0.11

This indicates the high adhesiveness of 3D-printed scaffolds, which, in turn, suggests high surface loyalty due to optimal roughness and hydrophilicity. It should also be noted that the 3D printing method itself provides additional surface relief, and due to the predicted porosity parameters, the total surface area receives significantly higher indicators as compared to the smooth polished surfaces of implants under the conditions of classical metal osteosynthesis. This phenomenon has a positive effect on improved adhesion values not only in the case of microbial cells but also in the case of all cell types, like for example in the case of osteoblasts.

As can be seen from Fig. 4, the total number of adhered and proliferated microbial cells on the surface of the control samples differs significantly. The activity of the functionalized surface of experimental scaffolds was recorded, namely the appearance of additional biological properties in the form of a bactericidal and bacteriostatic effect.

The presence of AgNPs on the surface of the samples prevented significant bacterial growth compared to the controlled samples after 24 h of culturing microbial cells under optimal conditions for proliferation.

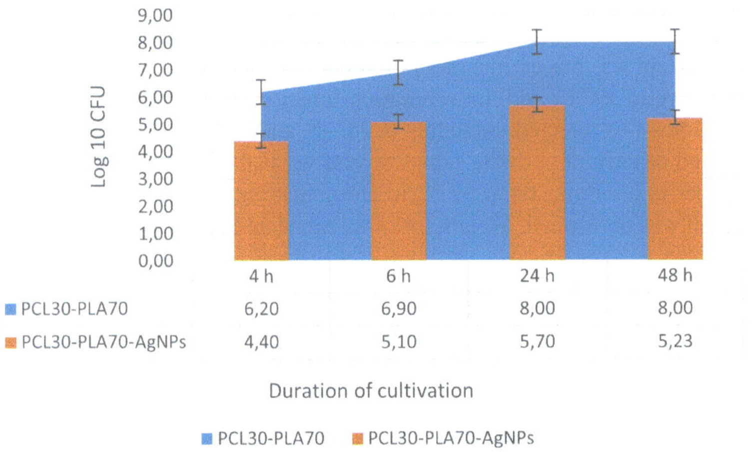

Fig. 4. Intensity of adhesion and proliferation of microbial cells on the surface of scaffolds.

At the same time, the final number of S. aureus did not exceed 6 log CFU even after 48 h of observation. It is known that AgNPs can release Ag + ions, which disrupt bacterial function through reaction with the cell membrane [21]. It should be noted that after 48 h of observation, there is a decrease in the total number of adhered and proliferated microbial cells on the surface of experimental samples as compared to the one after the made control. This phenomenon can be explained by the additional antimicrobial effect of the material with silver nanoparticles due to the onset of material degradation and the release of active nanoparticles into the intercellular fluid with the subsequent release of Ag + ions. After all, hydrolytic chain cleavage is the most common method of degrading high molecular-weight polyesters. There are two main types of it: bulk or surface degradation. The first leads to a decrease in the molecular weight of the polymer due to the release of by-products of the terminal carboxyl and hydroxyl groups, and the second leads to partial destruction of the structure of the fibers themselves. Continuous ion release can protect the implant surface from bacterial contamination after surgery and avoid bacterial contamination in the postoperative period after implantation. At the same time, the molecular weight of the material does not change significantly during surface destruction since surface by-products leave the surface, diffusing into the environment, which leads to the liquefaction of the material.

It is interesting to compare the numerical indicators of the number of adhered and proliferated microbial cells on the functionalized surface (Fig. 4) and the data on the mass loss of the studied samples during material degradation in the SBF solution (Fig. 5). This experiment was conducted in parallel with the framework of the hypothesis regarding additional properties of the functionalized surface of the degrading material.

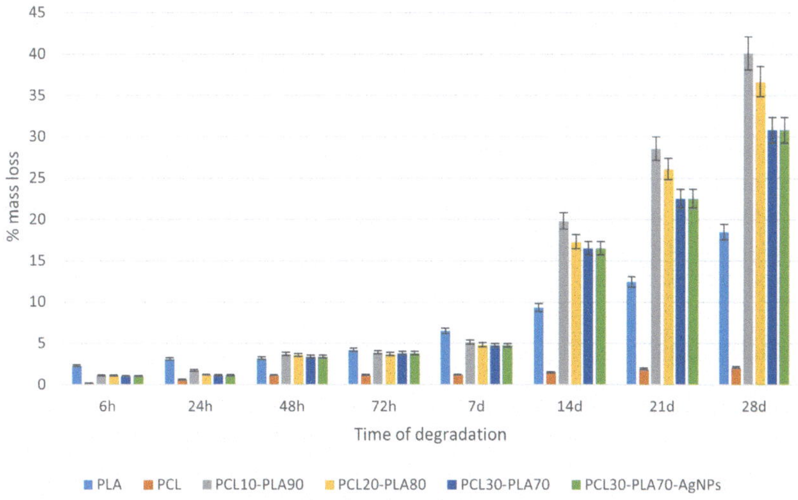

Fig. 5. Degradation of copolymer scaffolds in SBF solution.

As can be seen from Fig. 6, under the conditions of comparing the degradation indicators and quantitative reflection of microbial cell proliferation, a dependence curve is observed during the final observation period. The last observation period (48 h) of microbial cell cultivation coincides with the beginning of degradation of the experimental material, even with a visual assessment of the mass loss graph (Fig. 5).

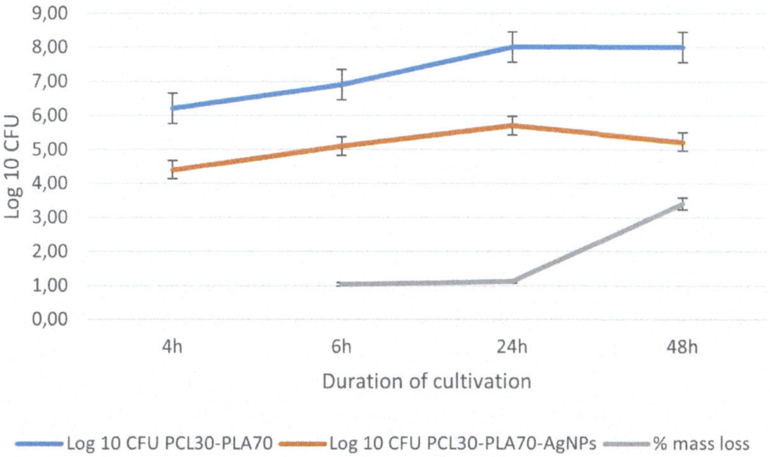

Fig. 6. Dependence of the number of adhered cells and mass loss during degradation.

Importantly, the degradation behavior can be actively controlled by adjusting the ratio of PCL and PLA in the polymer blends and by varying the filling percentage of the 3D framework. Just by controlling the amount of PCL, the degradation rate can increase to 20% or more [24]. If porosity during 3D printing and the overall geometry of the framework are considered, degradation rates can be further on optimized. It should also be noted that the degradation rate in SBF solution is significantly affected by the surface wettability index. The level of hydrophilicity for pure PCL and PLA is completely different.

Thus, when predicting the properties of the final polymer, it is necessary to consider the mathematical relationship and wettability intensity of each component and to consider the total surface area of the experimental framework, which depends on the thickness of the printed layer and the percentage of the porosity. Further research will be devoted to these issues.

5 Conclusions

This study successfully demonstrated the antibacterial properties of 3D-printed biodegradable scaffolds composed of PLA and PCL with silver nanoparticles. The incorporation of AgNPs significantly reduced bacterial adhesion and proliferation on the scaffold surfaces, for example, against *Staphylococcus aureus*. Specifically, bacterial density on PCL30-PLA70-AgNPs was 5.70 ± 0.13 log CFU at 24 h and 5.23 ± 0.11 log CFU at 48 h, compared to 8.00 ± 0.14 log CFU and 8.00 ± 0.16 log CFU, respectively, on PCL30-PLA70 without AgNPs. These findings confirm that the presence of AgNPs creates the antibacterial potential of the scaffolds while maintaining biocompatibility.

The activity of the functionalized surface of the experimental scaffolds in the form of a bactericidal and bacteriostatic effect was proven. The observed presence of silver nanoparticles on the surface of the samples prevented significant bacterial growth, as compared to the controlled samples, after 24 h of cultivation of microbial cells in optimal conditions for reproduction. This phenomenon can be explained by the additional antimicrobial effect of the material with silver nanoparticles due to the onset of material degradation and due to the release of active nanoparticles into the intercellular fluid with the subsequent release of silver ions. Within the framework of the hypothesis regarding additional properties of the functionalized surface of the degrading material, the dependences of degradation indicators and quantitative reflection of microbial cell proliferation were obtained at the final observation period (48 h) of microbial cell cultivation, which coincides with the beginning of degradation of the experimental material.

The novelty of the research lies in the development of a functionalized 3D-printed polymer skeleton with controlled biodegradability and long-lasting antibacterial action. Unlike traditional metallic implants, these scaffolds offer tunable mechanical properties, improved biocompatibility, and environmentally friendly degradation. Adjusting the PCL/PLA ratio allows for controlled degradation while maintaining structural integrity, making the material adaptable for various biomedical applications.

Future studies will be focused on optimizing 3D-printing parameters, such as porosity and surface modifications, which can enhance the structural properties of material and degradation behavior for advanced tissue engineering applications. Also, it will be planned to explore the effects of different AgNP concentrations on osteointegration, mechanical properties, corrosion resistance, hydrophilicity, and cytotoxicity.

Acknowledgment. This research was funded by grants from the Ministry of Education and Science of Ukraine (№ 0125U000447 and 0125U001429). The research was funded by the EU

NextGenerationEU through the Recovery and Resilience Plan for Slovakia under the project No. 09I03-03-V01-00101 and by the projects VEGA 1/0453/24, KEGA 012TUKE-4/2024 granted by the Ministry of Education, Research, Development and Youth of the Slovak Republic. The authors appreciate the support of the Internation-al Association for Technological Development and Innovations.

References

1. Armengol, ES., et al.: Unveiling the potential of biomaterials and their synergistic fusion in tissue engineering. Eur. J. Pharm. Sci. **196**, 106761 (2024). https://doi.org/10.1016/j.ejps.2024.106761
2. Nandhini, J., Karthikeyan, E., Rajeshkumar, S.: Eco-friendly bio-nanocomposites: pioneering sustainable biomedical advancements in engineering. Discover Nano **19**, 86 (2024). https://doi.org/10.1186/s11671-024-04007-7
3. Farjaminejad, S., Farjaminejad, R., Garcia-Godoy, F.: Nanoparticles in bone regeneration: a narrative review of current advances and future directions in tissue engineering. J. Funct. Biomater. **15**, 241 (2024). https://doi.org/10.3390/jfb15090241
4. Mahmood, A., Perveen, F., Chen, S., Akram, T., Irfan, A.: Polymer composites in 3D/4D printing: materials, advances, and prospects. Molecules **29**, 319 (2024). https://doi.org/10.3390/molecules29020319
5. Bhat, C., Prajapati, M.J., Kumar, A., Jeng, J.-Y.: Additive manufacturing-enabled advanced design and process strategies for multi-functional lattice structures. Materials **17**, 3398 (2024). https://doi.org/10.3390/ma17143398
6. Croes, M., et al.: Antibacterial and immunogenic behavior of silver coatings on additively manufactured porous titanium. Acta Biomater. **81**, 315–327 (2018). https://doi.org/10.1016/j.actbio.2018.09.051
7. Boo, G.-J.A., Grijpma, D.W., Moriarty, T.F., Richards, R.G., Eglin, D.: Antimicrobial delivery systems for local infection prophylaxis in orthopedic- and trauma surgery. Biomaterials **52**, 113–125 (2015). https://doi.org/10.1016/j.biomaterials.2015.02.020
8. Breisch, M., Grasmik, V., Loza, K., Pappert, K., Rostek, A., et al.: Bimetallic silver-platinum nanoparticles with combined osteo-promotive and antimicrobial activity. Nanotechnology **30**, 305101 (2019). https://doi.org/10.1088/1361-6528/ab172b
9. Zhang, R., Lee, P., Lui, V.C.H., Chen, Y., Liu, X., et al.: Silver nanoparticles promote osteogenesis of mesenchymal stem cells and improve bone fracture healing in osteogenesis mechanism mouse model. Nanomedicine **11**(8), 1949–1959 (2015). https://doi.org/10.1016/j.nano.2015.07.016
10. Myakinin, A., Turlybekuly, A., Pogrebnjak, A., Mirek, A., Bechelany, M., Liubchak, I., et al.: In vitro evaluation of electrochemically bioactivated Ti6Al4V 3D porous scaffolds. Mater Sci. Eng. C Mater. Biol. Appl. **121**, 111870 (2021). https://doi.org/10.1016/j.msec.2021.111870
11. Besinis, A., De Peralta, T., Handy, R.D.: The antibacterial effects of silver, titanium dioxide and silica dioxide nanoparticles compared to the dental disinfectant chlorhexidine on Streptococcus mutans using a suite of bioassays. Nanotoxicology **8**, 1–16 (2014). https://doi.org/10.3109/17435390.2012.742935
12. Thangadurai, M., Ajith, A., Budharaju, H., Sethuraman, S., Sundaramurthi, D.: Advances in electrospinning and 3D bioprinting strategies to enhance functional regeneration of skeletal muscle tissue. Biomater. Adv. **142**, 213135 (2022). https://doi.org/10.1016/j.bioadv.2022.213135213135

13. Hlushchenko, V., Ivakhniuk, T., Oleshko, T., Berladir, K., Smiyanov, V., Oleshko, O.: Modern approaches and possibilities of application of 3d modeling for tissue engineering and bone regeneration. Lit. Rev. East Ukr Med J. **11**(4), 337–351 (2023). https://doi.org/10.21272/eumj.2023;11(4):337-351
14. Budharaju, H., Zennifer, A., Sethuraman, S., Paul, A., Sundaramurthi, D.: Designer DNA biomolecules as a defined biomaterial for 3D bioprinting applications. Mater. Horiz. **9**, 1141–1166 (2022). https://doi.org/10.1039/D1MH01632F
15. Budharaju, H., Suresh, S., Sekar, M., P., De Vega, B., Sethuraman, S., et al.: Ceramic materials for 3D printing of biomimetic bone scaffolds – current state-of-the-art & future perspectives. Mater. Des. **231**, 112064 (2023). https://doi.org/10.1016/j.matdes.2023.112064
16. Budnik, AF., Rudenko, P.V., Berladir, KV., Budnik, OA: Structured nanoobjects of polytetrafluoroethylene composites. J. Nano- Electron. Phys. **7**(2), 02022 (2015)
17. Wei, S., Ma, J.-X., Xu, L., Gu, X.-S., Ma, X.-L.: Biodegradable materials for bone defect repair. Mil. Med. Res. **7**, 54 (2020). https://doi.org/10.1186/s40779-020-00280-6
18. Smola-Dmochowska, A., Lewicka, K., Macyk, A., Rychter, P., Pamuła, E., Dobrzyński, P.: Biodegradable polymers and polymer composites with antibacterial properties. Int. J. Mol. Sci. **24**(8), 7473 (2023). https://doi.org/10.3390/ijms24087473
19. Koch, F., Thaden, O., Conrad, S., Tröndle, K., Finkenzeller, G., Zengerle, R., et al.: Mechanical properties of polycaprolactone (PCL) scaffolds for hybrid 3Dbioprinting with alginategelatin hydrogel. J. Mech. Behav. Biomed. Mat. **130**, 105219 (2022). https://doi.org/10.1016/j.jmbbm.2022.105219
20. Xu, B.B., Ye, J., Fan, B.S., Wang, X.J., Zhang, J.Y., Song, S.T., et al.: Protein-spatiotemporal partition releasing gradient porous scaffolds and anti-inflammatory and antioxidant regulation remodel tissue engineered anisotropic meniscus. Bioact. Mat. **20**, 194–207 (2023). https://doi.org/10.1016/j.bioactmat.2022.05.019
21. Shebi, A., Lisa, S.: Evaluation of biocompatibility and bactericidal activity of hierarchically porous PLA-TiO2 nanocomposite films fabricated by breath-figure method. Mater. Chem. Phys. **230**, 308–318 (2019). https://doi.org/10.1016/j.matchemphys.2019.03.045
22. Augustine, R., Kalarikkal, N., Thomas, S.: Electrospun PCL membranes incorporated with biosynthesized silver nanoparticles as antibacterial wound dressings. Appl. Nanosci. Switz. **6**, 337–344 (2016). https://doi.org/10.1007/s13204-015-0439-1
23. Hye, J.H., Taekyung, Y., Woo-Sik, K., Sang, H.I.: Highly reproducible polyol synthesis for silver nanocubes. J. Cryst. Growth **469**, 48–53 (2017). https://doi.org/10.1016/j.jcrysgro.2016.09.038
24. Hlushchenko, V., Oleshko, T., Berladir, K., Oleshko, O., Rybarczyk, J., Boiko, V.: Study of the features of in vitro degradation of PCL, PLA, and polymer 3D scaffolds in static and dynamic conditions. In: Machado, J. et al. (ed.) Innovations in Mechanical Engineering III. ICIENG 2024, LNME. Springer, Cham (2024). https://doi.org/10.1007/978-3-031-62684-5_28

Ion Bombardment as Control Factor in Synthesis of 2D and 3D CuO Nanostructures

Andrii Breus(✉), Yelyzaveta Baranova, Yelyzaveta Zasid, and Maksym Lutsenko

National Aerospace University, 17, Vadima Manka Street, Kharkiv 61070, Ukraine
A.Breus@khai.edu

Abstract. Glow discharge plasma was implemented to reveal the effect caused by ions extracted from plasma and accelerated in DC plasma sheath with respect to the synthesis of copper oxide nanostructures of different dimensionality (2D and 3D). For the purpose, a copper cap with a small orifice in the upper part was employed to cover a part of a surface of a disk-shaped copper sample, while another part was left exposed to the ion fluxes accelerated across the voltage drop of 760 V. It was shown, that at the same temperature, the exposed area demonstrated a large yield of 2D and 3D nanostructures grown after 30 min in oxygen atmosphere maintained at the pressure of 360 Pa, while the surface treated with just neutral flux did not show any traces of nanostructures. Moreover, the intermediate case of the treatment with the flux where the active neutral plasma species were left after passing the plasma flux through the small aperture, while the most ions were removed, also resulted in the formation of 2D nanostructures yet at much lower yield than on the exposed to plasma area. The findings are important with respect to the development of a reliable, flexible, high-production and cost-effective technology of synthesis of metal oxide nanostructures of various morphology.

Keywords: Copper Oxide · Nanostructures · Plasma · Ion Energy

1 Introduction

For the last decade, copper oxide nanostructures have been extensively studied with a purpose of their employment as elements of sensors, supercapacitors, catalyst, water purification, and space exploration [1]. A lot of fabrication techniques have been developed with respect of achieving the different morphologies, including 1D (nanowires), 2D (nanoflakes, nanoleaves), and complex 3D (nanoflowers) nanostructure. Most of the techniques relate to the wet chemistry methods, when various liquid solutions are used with the following stirring, drying etc. procedures. Unfortunately, the production time usually takes more then 24 h for the chemical methods, and not all the morphology range is covered at that, since 1D nanostructure are much better grown by use of simple thermal methods. That is why significant efforts of various research groups are put on the development of alternative methods, which allows achieving the advantages such as whole range of the nanostructure dimensionality, large production yield, and flexibility of the production, while removing the drawbacks like the long fabrication time,

F. J. G. Silva et al. (Eds.): DSMIE 2025, LNME, pp. 15–22, 2025.
https://doi.org/10.1007/978-3-031-96413-8_2

use of hazardous chemicals or expensive equipment. Among the perspective methods, plasma-enhanced are considered as the most fit for the requirements, which resulted in implementation of almost all the known plasma discharges such as glow, arc, magnetron, arc, radiofrequency (RF), and microwave (MW) to the problem of synthesis of copper oxide nanostructures. Unfortunately, for now the number of the reports on the plasma technologies is much lower then for the chemical methods.

The research aim is to reveal the role of plasma in synthesis of CuO nanostructures, namely, the effect that causes the bombardment of a surface where nanostructures grow, by ions extracted from plasma and accelerated in plasma sheath up to the energies of a few hundred electron-volts (eV). For the purpose, a part of the surface of the samples that underwent the action of plasma, was exposed to the direct plasma action, while another part was covered to track the effect the flux of neutral plasma species.

2 Literature Review

3D flowerlike CuO nanostructures made of interconnected nanowires were successfully synthesized in a ternary microemulsion, span-80/water/paraffin liquid at 80 °C for 0.5–8 h [2]. CuO nanostrips, nanowires and microspheres were obtained by using a wet chemistry method with a production time of a few hours [3], to be tested for sensing applications. Copper foam was engaged to grow 2D flower-like CuO nanostructures in a liquid solution, with the time of the process of about 3 h [4]. The diameter of the nanoflowers reached the values up to 5 μm, with the thickness of the substructures that compose the nanoflower, of about 50 nm. Due to the interaction between CuO and Cu foam, a high capacitance, good rate capability and cyclability was reached for the composite. Sonication method was applied for more than 3 h to create flower-like 3D CuO nanostructures doped with transition metals (Ni, Fe and Zn), with a purpose to study the optical properties as photocatalyst to degrade methylene blue dye [5]. Hedgehog-like nanostructures of CuO with sizes from 1 to 3μm, were successfully grown by a chemical reduction method [6]; however, the aging of the samples took the time from 30 to 60 days. The UV–vis spectra shown an outstanding degradation of organic dyes in the presence of the nanoparticles. CuO nanoleaves were fabricated and then covered with NiO nanoparticles to be used as sensitive and selective non-enzymatic nitrite sensors [7]. For the purpose, CuO seed layer was obtained in a hydrothermal process at 85 ∘C for the time ranged from 3 to 5 h. 3D binder-free CuO nanosheets-wrapped nanofilms has been in situ synthesized on Cu substrate by a simple and facile procedure, with an aim of fabricating high-performance glucose sensor [8].

3D porous CuO electrodes for supercapacitors were designed directly on Cu foam by the combination of femtosecond laser and electrochemical anodization, with the later step intended to reach the superior specific capacitance of the electrode and a high energy density [9]. For more than four decades, plasma has attracted a lot of attention as a surface treatment facility [10]. Various CuO nanostructures such as nanodots, nanowires, nanowalls, as well as "cabbage-like" 3D assemblies, were synthesized by direct oxidation of Cu thin films by micro-after glow atmospheric pressure microwave plasma ignited in Ar-O_2 mixture [11]. The time of the experiment was changed between 2 min and 8h. It was suggested that the type of the nanostructures can be controlled by choosing the appropriate experimental conditions with respect to the possible applications. The same research group performed a thorough study to investigate the role of stress in the formation of CuO 1D nanowires by micro-afterglow oxidation at atmospheric pressure [12]. It was established that, as for thermal oxidation, the nanowire length follows a parabolic dependence on time, which is described by transport limitations implied by the diffusion of species. Moreover, as for thermal oxidation, the growth of 1D nanostructures is guided by stress-induced grain boundary diffusion caused by the stress gradients, which require implementation of the methods of solving the stress and strain problems [13]. It was assumed, that the use of plasma enhances the rate of oxidation at the temperature much lower than used in pure thermal process, thus increasing the generation of stress and improving the conditions for synthesis of 2D nanostructures. Implementation of plasma as a production tool is explained by her extremely high chemical activity and ability to be controlled not only by gradients of gas pressure but electric and magnetic fields [14]. Moreover, the operation modes of plasma setups are considered as suitable as for deposition as for nanofabrication [15]. With respect to the production of the nanostructures, the investigation of mechanisms of the stress influence to the formation of copper oxide nanostructures confirmed their substantial effect as for the pure thermal oxidation, as in glow discharge plasma [16]. Radiofrequency plasma discharge proved to be a reliable tool in synthesis of 1D copper oxide nanowires, while more complex nanostructures (2D and 3D) were not obtained at that [17]. Arc discharge is also recognized as a proper source to generate very active plasma [18]. Successful application of arc discharge was reported with respect to synthesis of encapsulated (Fe-Cu-Al)@C structures that were grouped into flower-, ball-, tree-, and octopus-shaped structures with a substantial yield [19]. However, the most impressive results with respect to the flexibility of obtaining the rich and diverse nanostructures were found when employing glow discharge for the synthesis of nanostructures [20]. In should be stressed, that plasmas employed in all the sited reports, were quite different, and the effects of plasma parameters such as density, electron temperature, ion energy, exposition time, to mention but a few, should be studied.

3 Research Methodology

To carry out the experiments on the growth of copper oxide nanostructures, and reveal the effect caused by the ion bombardment, a vacuum chamber with a setup to ignite the glow discharge plasma was engaged. The electrodes were mounted in the cylindrical vessel with the diameter and height of 300 mm and 350 mm, respectively. The samples were installed on the cathode, and the schematic of the setup is shown in Fig. 1. For each experimental run (with ten runs as the total number to study the effect), a flat sample with a diameter of 25 mm and height of 2 mm was put on the cathode, and a cap with an outer diameter of 18 mm and a height of 9 mm was mounted on the sample (thickness of the walls was 2 mm). At that, only a peripheral part of the sample surface was exposed to the direct action of the glow plasma, while another part was screened by the cap. The cap has an orifice (1 mm diam.) in the center of the upper part, as well as the orifices along its edge. This structure allows decreasing significantly the ion flux from the total flux of particles extracted from the plasma to the surface of the sample covered by the cap. The distance between the anode and the flat surface of the cap was 16 mm. The voltage drop of 760 V allows reaching the current of 0.25 A at the pressure of oxygen maintained at 360 Pa. Each sample was treated for 30 min. During the plasma treatment, the sample was heated to the bright-red color, and its temperature was measured as 735–750° by use of a pyrometer.

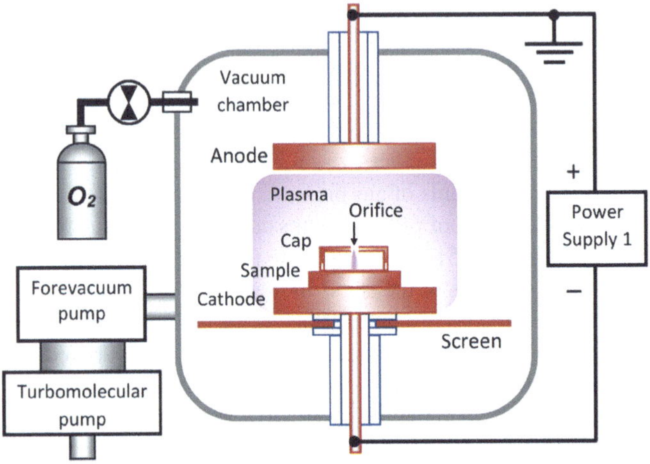

Fig. 1. A schematic of the experimental setup to grow copper oxide nanostructures in the glow discharge environment, when the copper sample is installed on the cathode, and part of the sample surface is screened by a cap with orifices to provide the oxygen delivery but remove the plasma component from the flux.

4 Results and Discussion

SEM images of different zones and nanostructures on the surface of the sample are shown in Fig. 2. The results illustrate the significant change in the nanostructure appearance along the sample surface, which can be associated with the action of high energy ions passed the high-voltage cathode drop. At the same time, the non-uniform ion treatment was accompanied with a uniform heating of the sample. Figure 2a illustrates the general view of 2D nanostructures of CuO developed on the surface of the sample right under the upper orifice in the cap (i.e. in the center of the sample), where the grown structure can be described as tree-like 2D nanostructure that was possibly synthesized as a sequence of 2D nanostructures occurring in the places of a damage of the previously grown structure. This effect can be promoted by a refined flux extracted from the plasma through the orifice, when the charged species (ions and electrons) are removed from the flux, while the neutral chemically-active species like radicals and exited molecules can pass through it. At the same time, other parts of the surface under the cap do not show any traces of nanostructures, as it is shown in Fig. 2b. The area outside the cap is much more interesting, since large amount of CuO nanostructures (mostly 2D) grown near the edges of the sample were observed there (Fig. 2c).

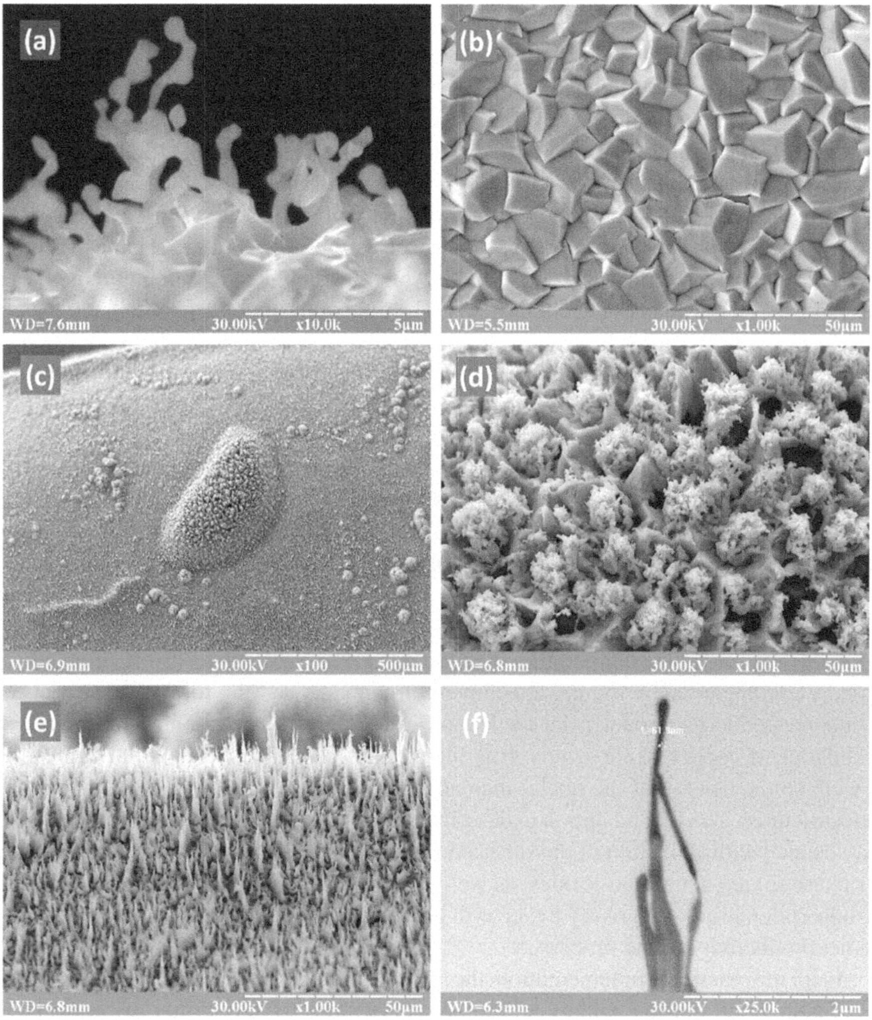

Fig. 2. SEM images of a sample treated on cathode and partially covered by a cap for 30 min in oxygen plasma (360 Pa, 760 V and 0.25 A): a – general view of 2D nanostructures of CuO developed on the surface of the sample right under the upper orifice of the cap (i.e. in the center of the sample); b – surface of the sample under the cap; c – CuO nanostructure (mostly 2D) grown near the edges of the sample outside the capped area; d – photograph revealing the transformation of CuO micro-grains into the nanostructures outside of the cap; e – magnified view of the field of 2D nanostructures (nanosheets) with height/width of 10 μm and thickness of 500 nm outside of the cap; f – nanostructure grown on top of the nanosheet with a diameter of 60 nm.

Intensive oxidation is the main driving force of their formation, and the photograph revealing the initial stage of transformation of CuO micro-grains into the nanostructures outside of the cap is shown in Fig. 2d. Magnified view of the field of 2D nanostructures (nanosheets) allowed measuring their height/width of 10 μm and thickness of 500 nm outside of the cap (Fig. 2e), while a fragment of a nanostructure grown on top of the nanosheet with a diameter of 60 nm is shown in Fig. 2f.

The obtained results allow concluding about the importance of ion bombardment as a driving force in obtaining the developed nanostructure pattern. In spite of the possible negative effect caused by ion sputtering to the growing structures, the total yield of the ion irradiation is positive with respect to promotion of change of morphology of the nanostructures, as well as in increase of productivity of the process. Moreover, the conducted experiments demonstrate the possibility of synthesis of 2D and 3D nanostructures on a solid sample, thus removing the necessity of collecting the yield in liquid media, as was reported on the most of the cited papers. The time of the process to obtain the typical sizes of the nanostructures (a few μm) was decreased from at least few hours to just half of hour. As for the possible ways of influence of the flux of energetic ions extracted from plasma and directed to a substrate, two options could be considered for the future theoretical and experimental studies. First, the effect of quick growth of oxide layers (CuO and Cu_2O) on Cu surface, when the stress in the layers does not relax and reaches the level of transition from one type of nanostructures to another (1D-2D-3D). After the analysis of the experimental results, it was concluded that the typical temperature increase rate is about 5 °C/min for in the thermal process of synthesis [1, 16, 17], which allows 0D nanodots to nucleate from CuO oxide grains and to continue their growth in the form of individual 1D nanowires. At the same time, the implementation of plasma heating leads to a sharp increase in the temperature increase rate [12, 17, 20], up to 150 °C/min, which is likely accompanied by mutual attachment of nanodots and the formation of roots of 2D copper oxide nanosheets; the nucleation centers could be presented by the active boundaries of copper oxide grains. It is likely that the stress promotes the nucleation as well, so the density of the nuclei increases, and 2D nanostructures start to form 3D nanostructures. The second option to describe the positive effect of the ion bombardment is associated with activation of the surface to promote the adsorption of species from the gas phase (oxide neutral molecules, as well as excited and charged molecules). In turn, the adsorption promotes greatly the growth of individual nanostructures, thus increasing the net productivity of the process.

As for the practical implementation, the results can be of great interest to the technologies dealing with sensing, energy and catalysis [1], since the possibility of controllable change of the morphology of the nanostructures is beneficial for the applications where the surface-to-volume ratio of material is critical.

5 Conclusions

The effect of ion bombardment to the synthesis of copper oxide nanostructures was investigated. Two main configurations of the ion flux extracted from glow plasma discharge and directed to the treated surface of copper sample were distinguished: (i) the flux of ions accelerated in the plasma sheath up to the energies of about 760 eV (by taking into account the typical charged states of ions in conventional glow discharges) by exposing the part of the surface to the glow plasma; (ii) the flux of active neutral plasma species separated from the plasma flux by a small orifice made in a central part of a cap that was used to cover another part of the sample surface. As a results, the yield of the nanostructures grown over the unprotected area was the most diverse as with respect to the morphology, since 2D and 3D nanostructures were obtained, as with respect to the phases of the growth. The effect of the neutralized plasma flux directed to the surface through the small orifice was also pronounced since 2D nanostructures were gown in that area. At the same time, the areas treated with just flux of neutral species did not exhibit any traces of the nanostructures. The results of the experiments allow concluding that the presence of charged particles is one of the main controls in the development of a process of synthesis, since their presence in the flux directed to the growing structure is vital for shaping the structures, when the same temperature in the growth area did not show such prominent effect.

Acknowledgment. Andrii Breus acknowledges the support from the project sponsored by the NATO Science for Peace and Security Programme under grant id. G5814 project NOOSE.

References

1. Baranov, O., et al.: Recent innovations in the technology and applications of low-dimensional CuO nanostructures for sensing, energy and catalysis. Nanoscale Horizons **8**, 568–602 (2023). https://doi.org/10.1039/D2NH00546H
2. Chu, D.-Q., Mao, B.-G., Wang. L.-M.: Microemulsion-based synthesis of hierarchical 3D flowerlike CuO nanostructures. Mater. Lett. **105**, 151–154 (2013). https://doi.org/10.1016/j.matlet.2013.04.067
3. Zhou, Q., Zhang, Y., Zeng, T., Wan, Q., Yang, N.: Morphology-dependent sensing performance of CuO nanomaterials. Anal. Chim. Acta **1171**, 338663 (2021). https://doi.org/10.1016/j.aca.2021.338663
4. He, D., Xing, S., Sun, B., Cai, H., Suo, H., Zhao, C.: Design and construction of three-dimensional flower-like CuO hierarchical nanostructures on copper foam for high performance supercapacitor. Electrochim. Acta **210**, 639–645 (2016). https://doi.org/10.1016/j.electacta.2016.05.196
5. George, A., et al.: Photocatalytic effect of CuO nanoparticles flower-like 3D nanostructures under visible light irradiation with the degradation of methylene blue (MB) dye for environmental application. Environ. Res. **203**, 111880 (2022). https://doi.org/10.1016/j.envres.2021.111880
6. Aguilar, M.S., Esparza, R., Rosas, G.: Time-dependent facile synthesis of CuO hedgehog-like nanostructures and their catalytic activity. J. Solid State Chem. **277**, 46–53 (2019). https://doi.org/10.1016/j.jssc.2019.05.034
7. Li, Y., et al.: Copper oxide nanoleaves covered with loose nickel oxide nanoparticles for sensitive and selective non-enzymatic nitrite sensors. Mater. Res. Bull. **149**, 111712 (2022). https://doi.org/10.1016/j.materresbull.2021.111712
8. Yao, W., Zhang, Y., Duan, T., Zhu, W., Yi, Z., Cui, X.: 3D hierarchical walnut-like CuO nanostructures: Preparation, characterization and their efficient catalytic activity for CO oxidation. Physica B **493**, 7–13 (2016). https://doi.org/10.1016/j.physb.2016.03.028
9. Wang, S., et al.: High-performance 3D CuO/Cu flowers supercapacitor electrodes by femtosecond laser enhanced electrochemical anodization. Electrochim. Acta **293**, 273–282 (2019). https://doi.org/10.1016/j.electacta.2018.09.144
10. Levchenko, I., Romanov, M., Baranov, O., Keidar, M.: Ion deposition in a crossed E×B field system with vacuum arc plasma sources. Vacuum **72**, 335 (2004). https://doi.org/10.1016/j.vacuum.2003.09.002

11. Altaweel, A., Filipič, G., Gries, T., Belmonte, T.: Controlled growth of copper oxide nanostructures by atmospheric pressure micro-afterglow. J. Cryst. Growth **407**, 17–24 (2014). https://doi.org/10.1016/j.jcrysgro.2014.08.029
12. Altaweel, A., Gries, T., Migot, S., Boulet, P., Mézin, A., Belmonte, T.: Localised growth of CuO nanowires by micro-afterglow oxidation at atmospheric pressure: Investigation of the role of stress. Surf. Coat. Technol. **305**, 254–263 (2016). https://doi.org/10.1016/j.surfcoat.2016.08.001
13. Gontarovskyi, P., Smetankina, N., Garmash, N., Melezhyk, I:. Numerical analysis of stress-strain state of fuel tanks of launch vehicles in 3D formulation. In: Nechyporuk, M., Pavlikov, V., Kritskiy, D. (eds.) Integrated Computer Technologies in Mechanical Engineering – 2020, ICTM 2020, LNNS, vol. 188. Springer, Cham (2021). https://doi.org/10.1007/978-3-030-66717-7_52
14. Baranov, O., Romanov, M., Fang, J., Cvelbar, U., Ostrikov, K.: Control of ion density distribution by magnetic traps for plasma electrons. J. Appl. Phys. **112**(7), 073302 (2012). https://doi.org/10.1063/1.4757022
15. Baranov, O., Romanov, M., Wolter, M., Kumar, S., Zhong, X., Ostrikov, K.: Low-pressure planar magnetron discharge for surface deposition and nanofabrication. Phys. Plasmas **17**, 053509 (2010). https://doi.org/10.1063/1.3431098
16. Breus, A., Abashin, S., Serdiuk, O.: Control of stress conditions in growth of copper oxide nanostructures. In: Nechyporuk, M., Pavlikov, V., Kritskiy, D. (eds.) Integrated Computer Technologies in Mechanical Engineering - 2022. ICTM 2022. LNNS, vol. 657, pp. 11–122. Springer, Cham (2023). https://doi.org/10.1007/978-3-031-36201-9_10
17. Filipic, G., Baranov, O., Mozetic, M., Ostrikov, K., Cvelbar, U.: Uniform surface growth of copper oxide nanowires in radiofrequency plasma discharge and limiting factors. Phys. Plasmas **21**, 113506 (2014). https://doi.org/10.1063/1.4901813
18. Sysoiev, I., Shyrokyi, Y., Fesenko, K.: Long-life pulsed vacuum-arc plasma source. Probl. Atom. Sci. Tech. **2023**(5), 152–157 (2023). https://doi.org/10.46813/2023-147-152
19. Breus, A., Abashin, S., Serdiuk, O., Sysoiev, I.: Synthesis of metallic alloy particles on flat graphitic interfaces in arc discharge. Arch. Mater. Sci. Eng. **121**(2), 49–59 (2023). https://doi.org/10.5604/01.3001.0053.8486
20. Breus, A., Abashin, S., Serdiuk, O.: Formation of 2D carbon nanosheets and carbon-shelled copper nanoparticles in glow discharge. In: Advanced Manufacturing Processes V. InterPartner 2023, LNME, Springer, Cham, pp. 256–264 (2024). https://doi.org/10.1007/978-3-031-42778-7_23

Resistance to Wear During Friction with Boundary Lubrication of Cast Iron-Iron Pairing with Nanocrystalline Structure-Reinforced Surface Layers

Volodymyr Gurey[1](✉), Ihor Hurey[1], Marian Bartoszuk[2], Yurii Kholod[1], and Rostyslav Hera[1]

[1] Lviv Polytechnic National University, 12, S. Bandery Street, Lviv 79013, Ukraine
Volodymyr.I.Hurei@lpnu.ua

[2] Opole University of Technology, 5, Mikolajczyka Street, 45-271 Opole, Poland

Abstract. Thermo-deformation treatment (TDT) is a surface-strengthening method that employs concentrated energy sources. In the contact area between the tool and the treated surface, the surface layer of a metal is heated rapidly to reach phase transition temperatures. This heating is followed by a quick cooling process, resulting in the formation of strengthened nanocrystalline layers. Research has demonstrated that a strengthened nanocrystalline layer is formed in the surface layer of cast iron (EN-GJL-200) parts during TDT. This layer has a thickness of 90–120 μm and a microhardness of 7–8 GPa. Near the treated surface, the grain size of the strengthened layer measures 20–40 nm. This strengthened layer significantly enhances the performance of a pair of the "grey cast iron-grey cast iron" during sliding friction with boundary lubrication. In the process of friction with boundary lubrication, when the unit load is increased above 2 MPa, the unstrengthened friction pair works for 2–3 h, and then catastrophic failure begins. Strengthened specimens operate normally when the unit load is increased to 6 MPa. Specimens strengthened using a tool with transverse grooves on the working part have a lower wear rate (by 30–40%) compared to those strengthened using a tool with a smooth work part. Notably, only one component of the friction pair was subjected to strengthening.

Keyword: White layer · Thermo-deformation Treatment · Strengthening

1 Introduction

The reliability of technological products is fundamentally rooted in the quality of the single parts that comprise them. This quality is unequivocally determined by their operational properties. They are influenced crucially by the physical, chemical and mechanical characteristics of the surface layer, along with the quality parameters of the working surfaces produced during the finishing treatment [1]. The treated surface and surface layer's quality directly impacts the operational capabilities of machine parts [2, 3], specifically

F. J. G. Silva et al. (Eds.): DSMIE 2025, LNME, pp. 23–32, 2025.
https://doi.org/10.1007/978-3-031-96413-8_3

their resistance to fatigue fracture [4, 5], wear [6], and corrosion resistance [7]. It is imperative to recognize that all fracture processes initiating during product operation start at the surface.

The surface layer of machine parts inevitably wears down due to sliding friction between contacting surfaces. Several critical factors significantly impact the interaction between these surfaces, including the actual contact area, local stress, contact stiffness, and the formation conditions of the oil film [3]. The wear resistance and contact strength of surface layers are primarily determined by essential quality parameters such as layer thickness, microhardness, residual stresses (both magnitude and sign), as well as structure, texture, and grain size. Understanding and optimizing these factors is crucial for enhancing the performance and longevity of machine components [6].

The main structure of this paper is organized as follows. The second section provides a literature review on surface treatment methods and the formation of strengthened layers with a nanocrystalline structure. The third section describes the TDT method, in which strengthened surface layers with a nanocrystalline structure are formed, and the methodology for studying the wear resistance of friction pairs. The fourth section discusses the results of experimental studies of the wear resistance of grey cast iron specimens with strengthened surface layers. Some conclusions are drawn in the fifth section.

2 Literature Review

Advanced surface treatment methods for machine parts are essential for significantly enhancing the operational properties and performance of components and products [9]. These methods enable the control of surface characteristics such as waviness, roughness, and bearing capacity, along with key traits like hardness, residual stresses, structure, phase and chemical composition, and texture in the surface layer of machine parts. By implementing these treatments, it is possible to effectively improve friction resistance, reduce wear, and combat fatigue and corrosion, ultimately improving the durability and reliability of the components.

To significantly enhance the operational properties of machine parts, it is essential to strengthen the surface layers of their contacting surfaces [1, 3]. However, creating strengthened surface layers on the working surfaces of these parts while maintaining the required accuracy and quality poses a significant challenge. Various methods are employed to achieve high levels of accuracy and quality on treated surfaces. These methods include the formation of a regular profile on contacting surfaces, grinding with different types of abrasive wheels, employing SPD [8], applying various coatings [9–12], surfacing with alloy metals, and utilizing electrophysical and electro-chemical processing techniques [13], among others.

The various methods using concentrated energy sources [9] are employed, including laser [14], plasma, ion-beam [15], and friction treatments to reinforce the surface layers of machine parts that operate under friction. These processes rapidly heat localized areas of the metal to temperatures above its phase transformation point, followed by quick cooling after the heat source is removed. This results in strengthened layers, or white layers, with a nanocrystalline structure (NCS) that possess unique physical and mechanical properties dissimilar with the base metal [16, 17]. These enhanced layers significantly improve the performance of machine parts.

Methods involving intensive plastic deformation are used to grind metal grains that promote increased hardness and modify accordingly their properties [16, 17]. These methods result in the grinding of grains and the creation of metal with a nanocrystalline structure (NCS) [18]. Metals with the NCS undoubtedly demonstrate significantly enhanced operational properties.

Thermo-deformation treatment (TDT) is a method for strengthening machine part surfaces. It combines two methods: employment of a concentrated energy source and intense plastic deformation of the metal's surface layer of the treated areas. This energy source is generated during high-speed friction (60–90 m/s) in the contact zone between the tool and the treated surface. The temperature rises rapidly ($5 \cdot 10^5$ – $4 \cdot 10^6$ K/s) to temperature exceeding the phase transition point of the metal in this zone. Parallelly the shear deformation occurs in the surface layer of the metal. The tool used in this process features transverse grooves on its periphery. This design interrupts the flow of thermal energy and enhances the intensity of shear deformation. As the contact zone moves along the treated surface, the local area of the metal surface layer cools quickly ($2 \cdot 10^5$ and $1 \cdot 10^6$ K/s). The outcome of this process is the formation of strengthened surface layers with NCS (white layers) on the machine parts.

The study aimed to investigate the influence of the quality of the strengthened layer on the wear resistance at sliding friction with boundary lubrication of the grey cast iron EN-GJL-200 – Grey cast iron EN-GJL-200 pair.

3 Research Methodology

The study examined wear resistance under sliding friction with boundary lubrication, conducted using the testing machine UMT-1 following a scheme 'Ring-Ring' (Fig. 1). The unit load was adjusted within a 0.2 MPa to 1 MPa range, while the sliding speed varied from 0.5 m/s to 1.8 m/s. Each test stage lasted for 110 h. The oil "Industrial-30A" was delivered to the friction zone at 15 to 20 drops per minute.

The stationary specimen was connected to a differential force meter sensor. The signal from this sensor was recorded by an analog-to-digital converter from National Instruments, capturing the friction torque used to determine the friction coefficient. Additionally, thermometric transducers with thermoelectrodes, each with a diameter of 0.2 mm, were mounted on the specimens. The signals from these transducers were also sent to an analog-to-digital converter, which recorded the integrated temperature at a depth of 0.3 to 0.5 mm from the friction surface.

Before the test, all friction pairs were run-in to stabilize the friction torque and ensure proper adhesion between the mating surfaces. This was evaluated by checking for friction marks covering at least 90% of the working friction surface on each specimen.

The TDT operates similarly to the grinding process, particularly in terms of kinematics and execution (Fig. 2). It is executed using a surface grinding machine-tool, where a specialized metal tool (disk) with a diameter matching that of the abrasive wheel is installed instead of the standard abrasive wheel. To effectively generate a concentrated heat flux, it is essential to achieve a linear velocity of 60–90 m/s at the periphery of the tool's contact with the workpiece. In comparison, the typical grinding wheel velocity on these machine tools is only 36 m/s. Consequently, upgrading the main drive unit of the grinding wheel is imperative to enhance the tool's velocity [19].

The TDT was performed on a modernized KNUTH surface grinding machine, model HFS 3063 VS (Germany). A frequency converter is installed to achieve the required speed. The outer diameter of the disk is 360 mm, and the width of the tool's working part is 6 mm. A metal disk was mounted on the faceplate, a standard component of this machine's equipment set and this assembly must maintain a static balance. The same balancing method used for abrasive tools is applied to balance the metal tool as well. More detailed processing process is described in papers [19–21].

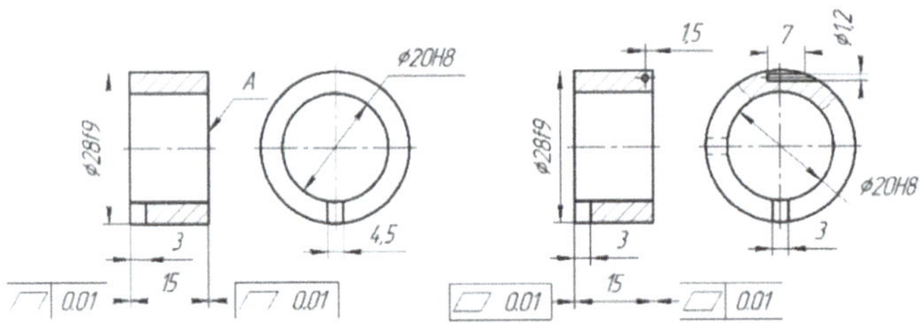

Fig. 1. Sketch of the specimen and counter-specimen for studies the wear during friction according to the scheme "Ring-Ring".

The wear criterion was based on the mass loss of specimens after 20 h of friction. This loss was measured using an analytical balance, which has an accuracy of ± 0.2 mg.

To enhance the intensity of shear deformation in the metal surface layer within the tool-part contact zone, a tool with transverse grooves was used (TTG) (see Fig. 2a). Additionally, a tool with a smooth working part (TSWP) was employed.

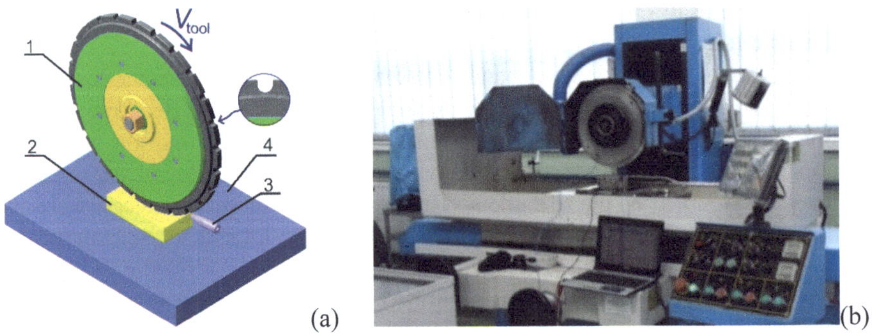

Fig. 2. Scheme of TDT of machine parts' flat surfaces (a): 1 – tool, 2 – workpiece, 3 – nozzle, 4 – machine-tool's table, V_{tool} – linear velocity of rotation of the tool; HFS 3063 VS surface grinder from KNUTH (Germany) (b).

The specimens of the friction pairs being studied were made of grey cast iron EN-GJL-200 after undergoing artificial ageing. The working surfaces of the moveable specimens were quench-hardened, while the fixed specimens were only ground. When both components of the friction pair are quench-hardened, the effect of increased wear resistance is diminished. An unstrengthened friction pair was used for comparison.

4 Results and Discussion

Experiments have demonstrated that TDT using a TSWP and a TTG significantly enhances the wear resistance of a cast iron-iron friction pair made of grey cast iron EN-GJL-200. When cast iron specimens strengthened with a TSWP were paired with unstrengthened (fixed) specimens under a unit load of 2 MPa, the wear rate of the moveable specimens decreased by almost five times, while that of the fixed specimens decreased by four times. In contrast, when the specimens were strengthened with a TTG, the wear rate of the moveable specimens was reduced by nearly eight times, and that of the fixed specimens decreased by 6.5 times compared to the unstrengthened friction pair. (Fig. 3).

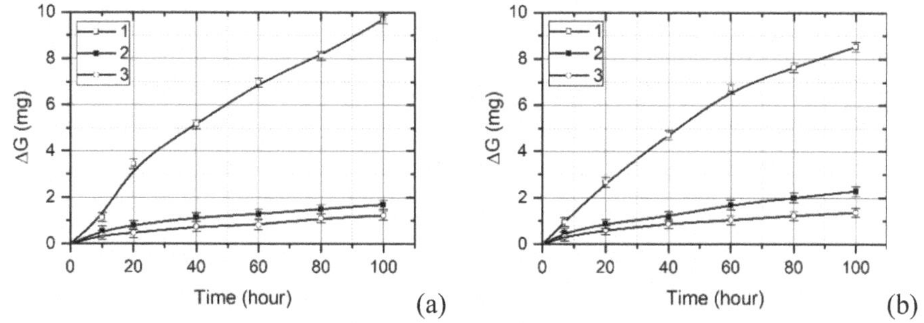

Fig. 3. Kinetic of wear of a pair of grey cast iron EN-GJL-200- grey cast iron EN-GJL-200 (a – moveable specimen, b – fixed specimen) (P = 2.0 MPa; V = 0,9 m/s): 1 – grinding; 2 – TDT using TSWP; 3 – TDT using TTG.

An increase in unit load leads to a higher wear rate. The wear intensity increases the most when the unit load reaches 0.75 MPa. However, further increasing the unit load up to 2 MPa results in a partial reduction in the wear rate. Under these friction conditions, the oxidative wear process mainly takes place. When the unit load exceeds 2 MPa, adhesion becomes the predominant type of wear for an unstrengthened ground pair. A sharp temperature rise is observed when friction occurs in an unstrengthened pair at a unit load of 4.5 MPa for 2–3 h. This leads to the destruction of the lubricating film, resulting in intense adhesion, the formation of scratches, and significant wear. Eventually, the friction pair becomes inoperable (Fig. 4). Studies on the wear intensity of strengthened pairs indicate that they operate normally with increasing unit load. A six-fold increase in unit load from 0.75 MPa to 4.5 MPa only leads to a 1.8–2.3 times increase in the wear rate of the strengthened specimens. In comparison, the wear rate for the unstrengthened specimens that work in conjunction with them increases by 2.6–3.3 times.

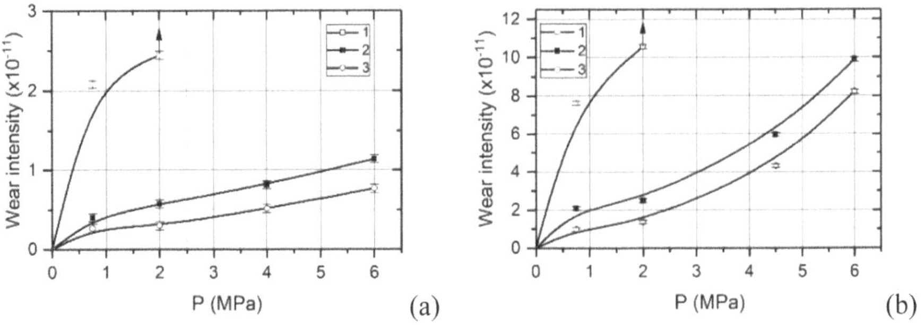

Fig. 4. Influence of unit load on wear of pair of grey cast iron EN-GJL-200- grey cast iron EN-GJL-200 (a – moveable specimen, b – fixed specimen) ($V = 0{,}9$ m/s): 1 – grinding; 2 – TDT using TSWP; 3 – TDT using TTG.

Experiments have demonstrated that the white layer significantly influences the temperature in the friction zone (Fig. 5). During the initial phase of friction, the temperature rises rapidly, reaching a peak value. This increase in temperature within the friction zone facilitates the formation of high-quality secondary structures on the friction surfaces of the components. These secondary structures enhance the contact area and decrease the actual unit load, resulting in a gradual reduction of temperature to the desired set point.

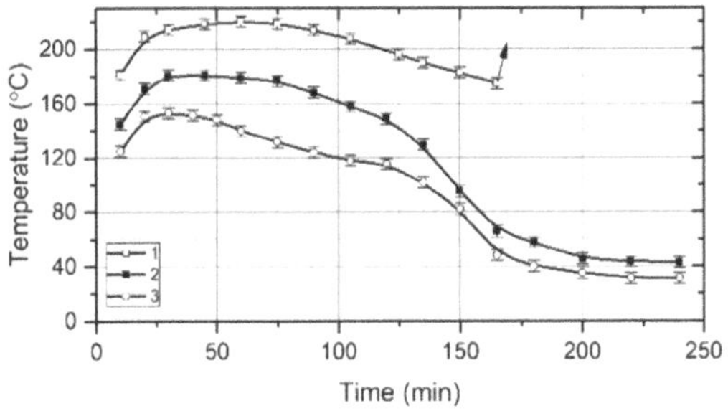

Fig. 5. Kinetic of temperature in the contact zone of a pair of grey cast iron EN-GJL-200- grey cast iron EN-GJL-200 ($P = 4.5$ MPa; $V = 0{,}9$ m/s): 1 – grinding; 2 – TDT using TSWP; 3 – TDT using TTG.

An increase in unit load results in a significant rise in temperature within the friction zone (see Fig. 6a). However, strengthened specimens exhibit a much smaller temperature increase as the unit load increases. When specimens are subjected to a load ranging from 0.75 to 4.5 MPa, a normal oxidative friction process occurs. If the unit load exceeds 4.5 MPa, there is a sharp increase in temperature.

Increasing the sliding speed also raises the temperature in the friction zone. Increasing the speed from 0.5 to 0.9 m/s does not significantly increase the temperature in the contact zone. However, when the sliding speed exceeds 0.9 m/s, the temperature in the friction zone increases rapidly. This indicates an intensive change in the phase and structural state of the metal in the friction zone. The minimum temperature in the friction zone at all the sliding speeds studied was recorded on specimens after TDT with TTG, and the maximum temperature was recorded on unstrengthened, ground specimens.

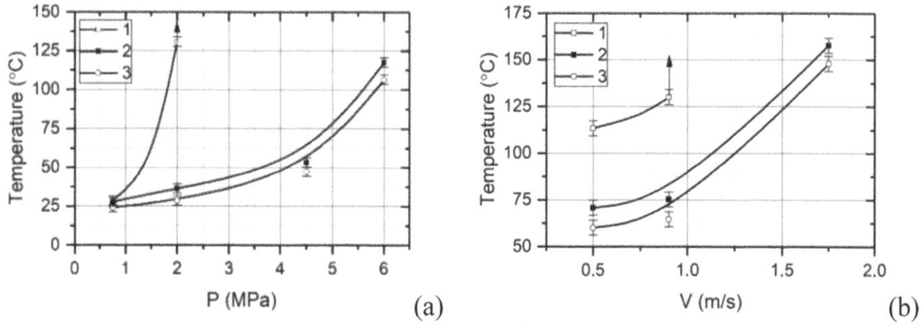

(a) (b)

Fig. 6. Influence of unit load ($V = 0.9$ MPa) (a) and sliding speed ($P = 4.5$ MPa) (b) on temperature in the contact area of the pair of grey cast iron EN-GJL-200 – grey cast iron EN-GJL-200: 1 – grinding; 2 – TDT using TSWP; 3 – TDT using TTG.

The TDT of the specimens significantly reduces the friction coefficient across all tested friction speeds and unit loads. Initially, the friction coefficient rises sharply, reaching its peak before declining rapidly to a stable value (Fig. 7a). During this process, favourable secondary structures form on the surface of the friction pair, enhancing its wear resistance. Additionally, the metal in the friction zone demonstrates structural adaptability.

When friction occurs in unstrengthened specimens, the friction coefficient increases with the unit load. This suggests that with an increase in the unit load, the setting process becomes more intense, micro-irregularities are broken off, and small solid particles are formed and when they enter the friction zone, act as abrasives and increase the amount of wear on the friction pair and the friction coefficient (Fig. 7b). However, after the rings undergo TDT, the value of the friction coefficient decreases. In this case, the main type of friction is oxidative. The friction surfaces of the specimens become smooth and shiny, with wear products being smeared across the working surface of the linings, and no adhesion is observed.

Metallographic studies have revealed that after TDT a nanocrystalline strengthened (white) is formed layer in the surface layers of the speciments. The shape of the tool's working surface significantly affects the quality parameters of the strengthened layer. The thickness of the white layer after TDT with TSWP was 90–110 μm, and with a TTG – 130–150 μm. The microhardness was 6.5 GPa and 8.6 GPa, compared to a microhardness of 2.0 GPa for the base metal. The grain size of the strengthened layer was 40–80 nm after TDT with a TSWP and 20–50 nm with a TTG. The density of dislocations in the white layer was between $(2.5–3.3)\cdot10^{11}$ cm^{-2}, in contrast to $(0.1–0.2)\cdot10^{11}$ cm^{-2} of the main structure. Additionally, the carbon content in the retained austenite reached 1.6 to 1.9%, compared to 0.8% in the main structure.

In boundary lubrication friction, secondary structures are formed on the friction surfaces, just as in non-lubricated friction. These consist mainly of iron oxides and alloying elements, as well as compounds of the active lubricant elements with metal

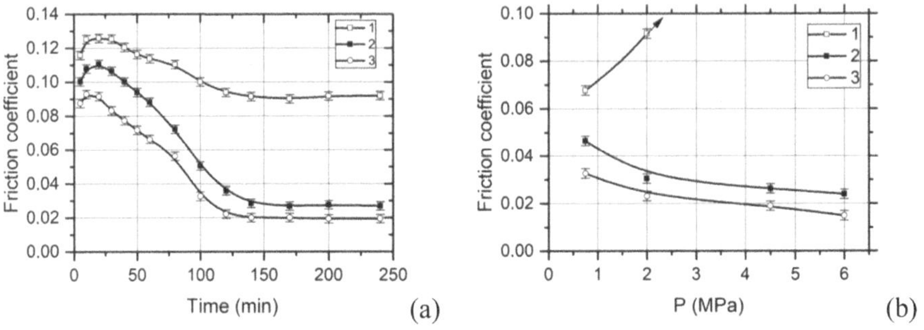

Fig. 7. Kinetic of friction coefficient ($P = 2.0$ MPa; $V = 0.9$ m/s) and influence of unit load ($V = 0.9$ m/s) on the friction coefficient of the pair of cast iron EN-GJL-200 – grey cast iron EN-GJL-200: 1 – grinding; 2 – TDT using TSWP; 3 – TDT using TTG.

atoms. The elements that form compounds with the base material can diffuse from the environment as well as from the metal volume (Fig. 8).

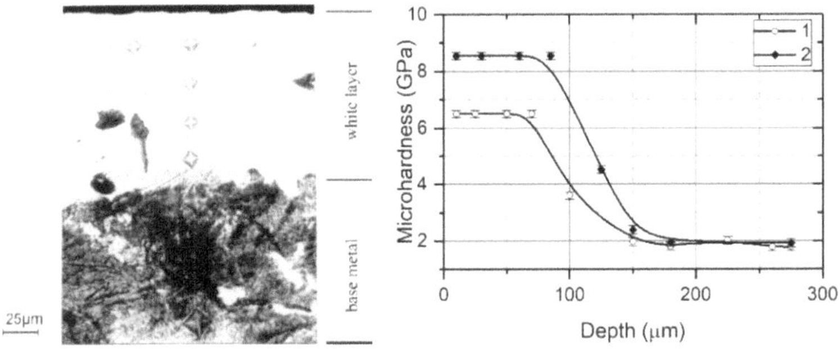

Fig. 8. Microstructure and microhardness of the strengthened layer obtained on grey cast iron EN-GJL-200.

Lubricants serve as a protective barrier that reduces the flow of oxygen into the friction zone, as their oxygen content is much lower than in air. In friction with a relatively low level of structural and thermal activation, this oxygen is sufficient to carry out complex tribochemical reactions. However, under severe friction conditions, the level of structural activation increases significantly. The amount of oxygen adsorbed on juvenile surfaces may not be sufficient for the formation of high-quality secondary structures [22]. Additionally, complex tribochemical reactions on friction surfaces consume energy stored in the form of dislocation energy. The increased density of dislocations activates the surface, increases the rate of diffusion and chemical reactions, and accelerates the formation of oxide films.

At contact areas, where significant plastic deformation and high local temperature occur, the lubricant film is destroyed, followed by adhesion of the juvenile surfaces with metal adhesion in microsites. Due to the mobility of lubricant molecules on friction surfaces, adsorption proceeds at a high speed, which allows restoring the integrity of the boundary film.

5 Conclusions

A strengthened nanocrystalline layer is formed on the surface during the TDT of cast-iron parts. This layer has a thickness of 90 to 120 μm and a microhardness of 7 to 8 GPa. The grain size of the strengthened surface layer is 20 and 40 nm near the treated surface. This strengthened layer significantly enhances the performance of a pair of the "grey cast iron-grey cast iron" during sliding friction with boundary lubrication. In the process of friction with boundary lubrication, when the unit load is increased above 2 MPa, the unstrengthened friction pair works for 2–3 h, and then catastrophic failure begins. Strengthened specimens operate normally when the unit load is increased to 6 MPa. Specimens strengthened with TTG have a lower wear rate (by 30–40%) compared to those strengthened with TSWP. It is important to note that only one component of the friction pair is strengthened. Once both components of the friction pair are strengthened, the increase in wear resistance ultimately diminished. In the future, we plan to study the wear resistance of strengthened specimens made from various steels during friction with boundary lubrication.

References

1. Dwivedi, D.K.: Surface engineering. Enhancing Life of Tribological Components. Springer India, New Delhi (2018). https://doi.org/10.1007/978-81-322-3779-2
2. Davis, J.R.: Surface engineering for corrosion and wear resistance. ASM International, Ohio (2001). https://doi.org/10.31399/asm.tb.secwr.9781627083157
3. Yushchenko, K.A., Borysov, Y., Kuznetsov, V.D., Korzh, V.M.: Surface Engineering. Naukova Dumka, Kyiv (2007)
4. Santecchia, E., et al.: A review on fatigue life prediction methods for metals. Adv. Mater. Sci. Eng. **2016**, 1–26 (2016). https://doi.org/10.1155/2016/9573524
5. Troshchenko, V.T., Khamaza, L.A.: Stages of fatigue failure of metals and alloys. Rep. Natl. Acad. Sci. Ukraine. **2**, 56–63 (2018). https://doi.org/10.15407/dopovidi2018.10.056
6. Hutchings, I., Shipway, P.: Tribology: Friction and Wear of Engineering Materials. Butterworth-Heinemann (2017)
7. Melters, P.M.: Introduction to Metal Fatigue. Aarhus University (2018)
8. Oborskyi, G., Orgiyan, A., Ivanov, V., Balaniuk, A., Pavlenko, I., Trojanowska, J.: Improvement of the dynamic quality of cantilever boring bars for fine boring. Machines **11**, 7 (2022). https://doi.org/10.3390/machines11010007
9. Ropyak, L., Shihab, T., Velychkovych, A., Dubei, O., Tutko, T., Bilinskyi, V.: Design of a two-layer Al–Al2O3 coating with an oxide layer formed by the plasma electrolytic oxidation of Al for the corrosion and wear protections of steel. Prog. Phys. Metals. **24**, 319–365 (2023). https://doi.org/10.15407/ufm.24.02.319
10. Kalchenko, V.V., Yeroshenko, A.M., Boyko, S.V., Ignatenko, P.L.: Development and research of thermoplastic methods for hardening details. Naukovyi Visnyk Natsionalnoho Hirnychoho Universytetu. 53–60 (2020). https://doi.org/10.33271/nvngu/2020-2/053
11. Kalchenko, V.V., Yeroshenko, A.M., Boyko, S.V., Ignatenko, P.L.: Determination of instantaneous temperature in the cutting zone during abrasive processing. Naukovyi Visnyk Natsionalnoho Hirnychoho Universytetu. (2019). https://doi.org/10.29202/nvngu/2019-5/14
12. Bazaluk, O., Dubei, O., Ropyak, L., Shovkoplias, M., Pryhorovska, T., Lozynskyi, V.: Strategy of compatible use of jet and plunger pump with chrome parts in oil well. Energies (Basel). **15**, 83 (2021). https://doi.org/10.3390/en15010083

13. Protsenko, V.S., Bobrova, L.S., Danilov, F.I.: Effects of water and sodium dodecyl sulfate additives on Cr(III) ions electroreduction in a deep eutectic solvent. Voprosy Khimii i Khimicheskoi Tekhnologii. 110–116 (2021). https://doi.org/10.32434/0321-4095-2021-135-2-110-116
14. Lesyk, D.A., Martinez, S., Mordyuk, B.N., Dzhemelinskyi, V.V., Lamikiz, A., Prokopenko, G.I.: Effects of laser heat treatment combined with ultrasonic impact treatment on the surface topography and hardness of carbon steel AISI 1045. Opt. Laser Technol. **111**, 424–438 (2019). https://doi.org/10.1016/j.optlastec.2018.09.030
15. Korzhyk, V., Tyurin, Y., Kolisnichenko, O.: Theory and practice of plasma-detonation technology of surface hardening metal products. Private Company Technology Center, Kharkiv, Ukraine (2021). https://doi.org/10.15587/978-617-7319-46-6
16. Saxena, A.K., Kumar, A., Herbig, M., Brinckmann, S., Dehm, G., Kirchlechner, C.: Micro fracture investigations of white etching layers. Mater. Des. **180**, 107892 (2019). https://doi.org/10.1016/j.matdes.2019.107892
17. Hosseini, S.B., Klement, U.: A descriptive phenomenological model for white layer formation in hard turning of AISI 52100 bearing steel. CIRP J. Manuf. Sci. Technol. **32**, 299–310 (2021). https://doi.org/10.1016/j.cirpj.2021.01.014
18. Shaw, L.L.: The surface deformation and mechanical behavior of nanostructured alloys. In: Nanostructured Metals and Alloys, pp. 481–506. Woodhead Publishing (2011). https://doi.org/10.1533/9780857091123.3.481
19. Hurey, I., et al.: Resistance to wear during friction without lubrication of steel-cast iron pairing with Nanocrystalline structure-reinforced surface layers. Lubricants. **11**, 418 (2023). https://doi.org/10.3390/lubricants11100418
20. Hurey, I., et al.: Influence of process liquids on the formation of strengthened nanocrystalline structures in surface layers of steel parts during thermo-deformation treatment. Appl. Sci. **14**, 8053 (2024). https://doi.org/10.3390/app14178053
21. Gurey, V., Maruschak, P., Hurey, I., Dzyura, V., Hurey, T., Wojtowicz, W.: Dynamic analysis of the thermo-deformation treatment process of flat surfaces of machine parts. J. Manuf. Mater. Process. **7**, 101 (2023). https://doi.org/10.3390/jmmp7030101
22. Tao, F.F.: A study of oxidation phenomena in corrosive wear. A S L E Trans. **12**, 97–105 (1969). https://doi.org/10.1080/05698196908972251

Two Layer Silane-Based Coatings for Paper Finishing

Nina Merezhko, Volodymyr Komakha(✉), Olga Komakha, and Victor Osyka

State University of Trade and Economics, 19, Kyoto Street, Kyiv 02156, Ukraine
v.komakha@knute.edu.ua

Abstract. The combined thin-layer protective coating for cellulose-containing materials includes a biocidal sublayer-adhesive composed of potassium methylsiliconate and an amino complex of copper and zinc methyl silicates, along with a basic protective layer of polyalkylhydridesiloxanes as the outer coating was studied. The synthesis of copper and zinc methyl silicates was carried out by reacting potassium methylsiliconate with copper sulfate and zinc chloride as fungicidal agents. An environmentally safe method for using copper and zinc methylsilicates through water-soluble ammonia complexes was suggested. The formation of chemical bonds between silanol groups and copper-zinc ions was demonstrated using IR spectroscopy. The use of two-layer organosilicon coatings has been substantiated to enhance moisture resistance and mold protection. The developed coatings can increase the hydrophobicity of the paper surface by up to 28.3% and reduce the hygroscopicity of paper by up to 20% compared to basic organosilicon products. Fungicidal properties and the ability to inhibit the development of mold are found on paper coated with adhesive sublayer from aqueous solution of the amino complex of copper and zinc methyl silicates and potassium methylsiliconate in a concentration of at least 1.5 and 2.0 wt.%, respectively.

Keywords: Paper Treatment · Wetting Angle · Biodamages Protection

1 Introduction

Paper is widely used across industries and economic sectors. The use of packaging and special-purpose paper with polymer composite coatings is rapidly expanding [1]. These materials contain a variety of ingredients: cellulose, synthetic and natural fibers, fillers, pigments, dyes and functional additives with relatively high chemical activity and the ability to interact with water [2]. Throughout its service life, paper is exposed to the destructive effects of the environment [3]. Therefore, there is a need for effective and economical ways to protect paper materials from damage caused by the action of moisture and microorganisms that reduce the service life and properties.

There are industrial products designed to inhibit the growth and reproduction of bacteria and fungi, but these materials have a number of disadvantages. In particular, the inability to maintain sufficient activity in conditions exposed to moisture and limited use in materials in contact with food [4].

F. J. G. Silva et al. (Eds.): DSMIE 2025, LNME, pp. 33–42, 2025.
https://doi.org/10.1007/978-3-031-96413-8_4

The study is based on the task of creating a protective coating, in which the use of new components and the method of their application will increase the resistance of the cellulose-containing material to external factors due to the formation of new bonds and a dense cross-linked composite structure of the coating, reduce water absorption, as well as provide comprehensive protection of paper from fungi biodamage.

2 Literature Review

The high chemical reactivity of paper components can lead to degradation over time. This is especially true for cellulose, which can break down when exposed to moisture and other environmental factors [5]. Water poses a significant threat to paper, as it can cause swelling, weakening, and eventual degradation of the paper fibers. This interaction may also result in the migration of fillers and pigments, thereby compromising the paper's appearance and structural integrity [6]. For the growth and development of microorganisms, as known, it is necessary that the level of humidity in their habitat is maintained within certain limits [7]. Water maintains normal osmotic pressure in the live cell, provides the possibility of metabolism [8], so almost all microorganisms are moisture-loving. To improve the water and moisture resistance of paper, a variety of materials can be utilized. Common synthetic coating agents include polyamideamine-epichlorohydrin, melamine formaldehyde, polyacrylamide, polyethyleneimine, polyvinylamine, and fluoropolymers. Natural wet-strength agents such as starch, chitosan, cellulose nanofibrils, and soy protein are also frequently used [9].

While natural coating materials, such as waxes, starch, cellulose nanofibrils and soy protein, often fall short in terms of performance [10], they offer a significant advantage in consumer safety, as they are mostly approved for contact with food products [11]. Synthetic polymer-based coatings offer improved barrier properties, enhanced durability and chemical resistance for paper materials. Polyethylene, polyethylene terephthalate, polybutylene terephthalate [12], polyvinylalcohol [13] coatings enhance the performance and range of applications of paper, but they do not always align with modern environmental strategies. However, coated paper is an appealing alternative to replace single-use plastic bags in packaging applications [14].

Fluoropolymer coatings offer superior water and oil repellency, excellent chemical resistance, and durability in extreme conditions [15, 16]. However, their high cost and environmental and health risks complicate their widespread use in the paper packaging industry [17].

Organosilicon coatings are increasingly used for treating cellulosic materials, such as textiles and paper, due to their advantageous properties [18, 19]. These coatings enhance water repellency [20], improve durability, and maintain the flexibility and mechanical strength of the materials [21, 22]. Additionally, they are environmentally friendly and compatible with organic compounds, making them suitable for various applications, including food packaging [23].

The task of comprehensive paper protection can be solved by using special coatings [24] combined with a complex of water-repellent and bactericidal and fungicidal protective properties [25], which will significantly increase the service life of cellulose-containing materials.

The selection of components for protecting cellulose-containing materials is based on several physico-chemical characteristics of paper substrates. These include the high content of reactive hydroxyl groups in cellulose macromolecules, which are present on nearly all surface areas [26]; increased water sorption capacity, leading to high hygroscopicity [27]; and low resistance to chemical compounds with non-neutral pH levels [28]. Considering these features and the nature of the interaction of silane coupling agents with cellulose [29].

The effective application of thin-layer organosilicon coatings depends on developing an optimal method for combining the high surface reactivity of cellulose-containing materials with siloxanes of different chemical compositions and structures. To achieve this, a scheme for applying a two-layer coating using methyl derivatives of potassium silicones and an amino complex of copper and zinc methyl silicates as a biocidal adhesive sublayer and oligoalkylhydridesiloxanes as a basic protective layer is proposed.

In this context, it is necessary to study both the synthesis processes of methyl-substituted copper and zinc silicates and their interaction with oligoalkylhydridesiloxanes.

3 Research Methodology

The study used industrial-grade electrical engineering paper (100% unbleached sulfate cellulose, 55% porosity, 60° water wetting edge angle). The protective coating comprised a biocidal adhesive sublayer of potassium methylsiliconate with copper and zinc amino complexes and an outer polyalkylhydridesiloxane layer.

Potassium methylsiliconate GKR-11 K was used in the form of a water-alcohol solution: density 1.17–1.21 g/cm^3, dry residue - 24–35 wt.%, silicon - not less than 4 wt.%, ethyl alcohol - 1.218 wt. %

Potassium methylsiliconate GKR-11 K was applied as a water-alcohol solution (density 1.17–1.21 g/cm3, dry residue 24–35 wt.%, silicon $\geq$ 4 wt.%, ethyl alcohol 1.218 wt.%). Industrial polyalkylhydridesiloxanes GKR 136–41 and GKR 94 M were used as aqueous emulsions (active hydrogen 1.30–1.80 wt.%, viscosity at 20 °C: 5–200 cSt). These siloxanes were chosen for their reactive groups, various hydrocarbon radicals, industrial availability, and non-toxicity.

The amino complex of copper and zinc methyl silicates was synthesized by reacting potassium methylsiliconate with copper sulfate and zinc chloride, forming insoluble metal methyl silicates and precipitated metal hydroxides via hydrolysis (1–3):

$$CH_3Si(OH)_2OK + H_2O \leftrightarrow CH_3Si(OH)_3 + KOH \quad (1)$$

$$xCH_3SI(OH)_2O^- + y\,Me_2+ \rightarrow \left[CH_3Si(OH)_2O\right]x\,Me_y \downarrow \quad (2)$$

$$Me_2^+ + 2OH^- \rightarrow Me\,(OH)^2 \downarrow, \left(Me = Cu^{2+}, Zn^{2+}\right) \quad (3)$$

Metal methyl silicate precipitates are solubilized via complexation with ammonia, forming aqueous amino complex solutions.

The biocidal sublayer-adhesive was prepared in aqueous solutions with potassium methylsiliconate (1.0–3.0 wt.%), copper and zinc methyl silicate amino complexes (1.0–2.0 wt.%), and distilled water (95.0–98.0 wt.%). The main protective layer consisted of a 2:1 mixture of polymethyl- and polyethylhydridesiloxanes as a 5–10% aqueous emulsion.

The coating process involved two stages with intermediate heat treatment: dipping the adhesive sublayer, drying at 120 °C for 2 h, followed by airless spraying of the main layer under pressure. The drying process ensured rapid solvent removal and coating stability.

Paper samples untreated, coated with the main protective layer (examples 1–3), and with the full organosilicon coating (examples 4–6) – were tested for water wetting angle, moisture absorption, and fungal resistance. Wetting angle was measured with an instrumental microscope ($\pm 1°$ accuracy). Fungal resistance was assessed after 30 days in a 29 $\pm$ 2 °C, 95–98% humidity chamber, rated from 0 (fungistatic) to 5 (high fungal growth). The FTIR spectra of the dispersed and tableted materials with KBr were recorded on a Specord IR-75 spectrometer in the range of 400–4000 cm^{-1}.

4 Results and Discussion

The reactions of interaction of potassium methylsiliconate with copper sulfate and zinc chloride were studied by direct potentiometric titration of metal salts with methylsiliconate solution while mixing the reaction medium. The most probable structure of these compounds is globules of branched methyl-substituted polysilicates of copper and zinc with particles of metal oxides and hydroxides evenly distributed between them. IR spectroscopic studies of the obtained methyl silicates also confirm the assumptions regarding the chemical interaction of silanol groups with copper and zinc ions to form metallosiloxane bonds (Fig. 1).

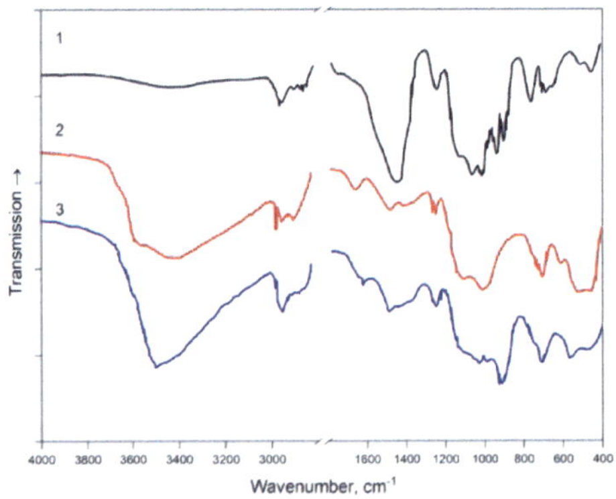

Fig. 1. IR spectra showcasing siloxane bond interactions in potassium methylsiliconate (1) and copper (2)/zinc (3) methyl silicates.

Wide intense absorption bands in the region of 3350 - 3600 cm^{-1}, characteristic of stretching vibrations of OH groups, were registered. Stripes at 3585 cm^{-1} and 3493 cm^{-1} were also noted respectively for copper and zinc methyl silicates, which probably belong to OH groups of metal hydroxides. There are absorption bands in the region of 2890 – 2975 cm^{-1}, characteristic of valence oscillations of CH in alkyl radicals. For copper and zinc methyl silicates, a significant decrease in the intensity of the band was recorded at 1425–1480 cm^{-1}, which belongs to anion carbonate, and at 750 cm^{-1}, which is responsible for symmetric deformation oscillations of the silicon-alkyl bond.

A characteristic of the copper and zinc methyl silicates is the disappearance of a narrow intense band at 890 cm^{-1}, which belongs to the anion carbonate, and a band at 925 cm^{-1}, which may be responsible for the oscillations of the Si–O–K bond. In addition, for zinc methyl silicate, absorption bands were noted at 970 and 890–900 cm^{-1}, which can be attributed to zinc hydroxide.

Siloxane bond of the obtained metal methyl silicates showed a significant decrease in the intensity of the bands at 1010–1020 and 1060 cm^{-1}, characteristic of three-dimensional siloxanes (1010–1020 cm^{-1}) and the branched siloxane chain (1056–1068 cm^{-1}). In addition, the strip at 1060 cm^{-1} shifts by 20 cm^{-1} into a lower-frequency region, which may indicate the presence of a chemical interaction of silanol groups with copper and zinc ions with the inclusion of metal atoms in the siloxane chain.

Displacement of absorption bands towards low frequencies during the replacement of potassium ions with copper and zinc was also recorded for the absorption band in the region of 500–690 cm^{-1}, assigned to the Si – O – Me group. Absorption bands were recorded for the Si – O – Cu groups at 610–625 cm^{-1}, and Si – O – Zn – at 550–595 cm^{-1}. Thus, the replacement of potassium ions with copper ions gives a shift in the Si-O-Me absorption band towards low frequencies by 50–75 cm^{-1}, and for zinc ions – by 126–136 cm^{-1} for methylsiliconates, which can serve as a confirmation of the interaction of Si-OH groups with copper and zinc ions to form polyalkylmetallosiloxanes.

Along with the presence of a complex of potential advantages of copper and zinc methylsilicates, they have a drawback due to their practical insolubility. It is proposed to transfer metal methyl silicate precipitates to a soluble state through the stage of complexation with ammonia. As a result of the interaction, complex cations $[Me(NH_3)_x]^{2+}$ are formed (where Me = Cu, Zn; x - is the coordination number, the predominant value of which is 4 for copper and 2 for zinc).

Comparison of IR spectra of powders obtained by drying at room temperature of aqueous solutions of metal methyl silicate amino complexes (Fig. 2) and their analogues obtained by direct synthesis (Fig. 1) showed that the intensity of the absorption band in the region of 3300–3600 cm^{-1} is slightly reduced, especially in the case of zinc-containing compounds.

Valence oscillations of the N-H bond characteristic of ammonia and ammonium ions (3030–3300 cm^{-1}) were not recorded, possibly due to the masking effect of a wide absorption band of OH groups. The intensity is also rather weak for the absorption band at 1380–1430 cm^{-1}, responsible for the deformation oscillations of the N-H bond. Copper and zinc compounds are characterized by the presence of very weak absorption bands at 1360 cm^{-1}, it is likely that during the drying of ammonia complexes, even at room temperature, the decomposition of complex cations $[Me(NH_3)x]^{2+}$ occurs with the

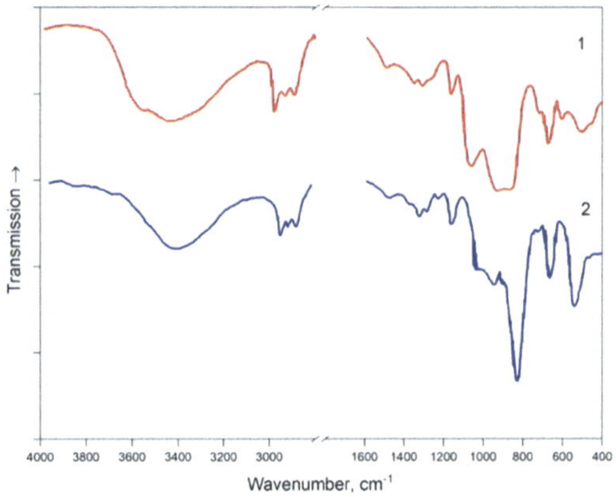

Fig. 2. IR spectra of copper (1) and zinc (2) methyl silicates obtained by drying amino complexes based on them.

removal of ammonia. Very weak absorption bands of 1345–1360 cm^{-1} may belong to N-H bonds of ammonia molecules sorbed by globules of polymethylmetalosiloxanes.

There are also certain changes in the nature of the absorption bands of the siloxane bond. In practice, they disappear completely at frequencies of 1060, 925 and 890 cm^{-1}, characteristic of the original potassium methylsiliconate. The shift towards low frequencies of the band at 1060 cm^{-1} for the latter in methyl silicates of metals obtained from amino complexes is even more noticeable, especially in copper compounds. The shear value reaches 45 cm^{-1} compared to 25 cm^{-1} for zinc methylsilicate (Fig. 2).

A significant decrease in the intensity of the absorption band of anion carbonate in the region of 1450–1470 and 750–780 cm^{-1} is characteristic of metal methylsilicates obtained through the stage of synthesis of amino complexes. The appearance of absorption bands responsible for deformation oscillations of groups was also noted: 625–630 cm^{-1} – Si–O–Cu and 555 cm^{-1} – Si–O–Zn.

Generalization of the obtained data allows us to conclude that when a mixture of methyl silicates and metal hydroxides is dissolved in aqueous ammonia, dissociation of Si – O – Me bonds occurs with the formation of Si – O – and complex cations $[Me(NH_3)_x]^{2+}$. Metal hydroxides dissolve in NH_4OH to form $[Me(NH_3)_x](OH)_2$, which dissociate in solution into complex metal cations and hydroxyl anions.

The results of the studies were the basis for the development of formulations of thin-layer fungicidal protective coatings containing copper and zinc methylsiliconates and organosilicon industrial hydrophobic agents – potassium methylsiliconate. The test results of paper samples exclusively with basic protective coatings (examples 1–3) and combined thin-layer organosilicon coatings according to example 4–6 and control (raw paper) are shown in Table 1.

The analysis of changes in the wetting angle of paper treated with a protective coating, in accordance with the proposed method, showed that the use of combined coatings can increase the contact angle of the paper surface by up to 28.3% and reduce the hygroscopicity of paper by up to 20% compared to basic organosilicon products. The study of the effect of microscopic fungi on paper treated with combined organosilicon

Table 1. Composition and properties of the paper samples

Example	Biocidal sublayer: mixture in aqueous solution wt.%		Protective layer: mixture of polyalkyl hydridesiloxanes in aqueous emulsion, wt.%	Paper properties		
	Potassium methylsiliconate	Amino complex of copper and zinc methyl silicates		Contact angle, deg, ± 1	Moisture absorption, wt.%	Fungal resistance, score
Raw paper	–	–	–	60	16.1	4
Example 1	–	–	5	77	12.9	3
Example 2	–	–	7	78	10.3	2
Example 3	–	–	10	85	9.2	1
Example 4	1	1	5	87	10.8	1
Example 5	2	1.5	7	89	8.1	0
Example 6	3	2	10	96	7.2	0

thin-layer coatings containing an amino complex of copper and zinc methyl silicates (at more than 1.5 wt.% in the adhesive sublayer) demonstrated that such coatings exhibit bio- and fungicidal properties and are capable of inhibiting the development of mold fungi. The combined use of potassium methylsiliconate and polyalkylhydridesiloxanes reduces the amount of adsorbed water, creating conditions unfavorable for the development of microorganisms on the paper surface.

Previous studies of two-layer tetraethoxysilane-polyethylhydridesiloxane coatings have shown the potential to achieve 84.5% screening of hydroxyl groups and a water wetting angle of 94° [30]. Given the similar hydrophobicity indices of the coatings proposed in this work, further research is needed to better understand and compare their interactions with cellulose-containing substrates. Expanding the applications of the developed coatings can be achieved by varying the composition of the main protective layer. For example, known coatings based on polydimethylsiloxane and hydrophobic silica nanoparticles have been shown to reduce boron release from wood samples and exhibit water contact angles of 114° and 172° [31]. Additionally, one-component coatings based on potassium methylsiliconate are known for their effectiveness in preserving aged papers and improving their mechanical properties [32]. The proposed fungicidal coatings could be valuable for preserving old prints and historical manuscripts. However, detailed long-term impact studies on paper substrates from different historical periods are needed.

Two-layer silane coatings enhance paper durability, improving water resistance, fungal protection, and longevity. They benefit packaging and archival use while reducing costs. However, optimizing chemical interactions, scalability, and application methods is essential for sustainability and effectiveness.

5 Conclusions

The research shows that it is possible to provide protection against biodamage and a stable degree of water repellency for cellulose-containing materials using two-layer coatings based on potassium methylsiliconates with copper-zinc methyl silicates as a biocidal sublayer and polyethylhydridesiloxane as an outer coating layer.

The possibility of forming chemical bonds involving free silanol groups of alkylsilicic acid derivatives and Si–H groups of polyethylhydridesiloxane has been established. In this combination of ingredients, a significant number of silanol groups in the sublayer can potentially enable chemisorption of the hydrophobizing agent layer, leading to the formation of strong siloxane bonds.

An environmentally safe method for using copper and zinc methyl silicates in water-soluble ammonia complexes with a concentration of 1.0–2.0 wt.% was proposed. Various options for incorporating copper and zinc methyl silicates at a level of 1.0–2.0 wt.% with potassium methylsiliconates in the sublayer composition, followed by an overlapping application of a 5–10% aqueous emulsion of polyalkylhydridesiloxanes, have been studied. The two-layer organosilicon coating increases moisture resistance by up to 2.2 times compared to untreated paper and by up to 27% compared to paper containing only the main polyalkylhydridesiloxane protective layer. The contact angle of the paper surface with water increased from 60° on untreated paper to 96° on paper treated with the two-layer organosilicon coating. The fungal resistance of paper treated with a two-layer organosilicon coating improved from a score of 4 (indicating active fungal growth on the nutrient substrate) to a score of 0 (indicating no fungal growth).

The obtained results may be of practical interest for the paper protection and the development of packaging materials based on it. Further research is needed to optimize the drying kinetics of the adhesive sublayer and the main protective layer, as these factors influence the structure, mechanical properties, and protective effectiveness of the coating.

References

1. Osyka, V., Merezhko, N., Koptjukh, L., Komakha, V., Kniaz, S.: Polyamidaminepichlorohydrin application for eco-safe waterfast and waterproof food packaging paper production. J. Food Sci. Technol.-Ukraine **15**(2), 112–122 (2021). https://doi.org/10.15673/fst.v15i2.2102
2. Ciolacu, D.: Cellulose allomorphs: structure, accessibility and reactivity. Polymer Science and Technology Series, Nova Science Publisher, Inc., New York (2010). ISBN 978-1-61668-323-8
3. Hubbe, M.: Paper's resistance to wetting - a review of internal sizing chemicals and their effects. BioResources **2**(1), 106–145 (2007)

4. Hubbe, M., Gardner, D., Shen, W.: Contact angles and wettability of cellulosic surfaces: a review of proposed mechanisms and test strategies. BioResources **10**(4), 8657–8749 (2015). https://doi.org/10.15376/biores.10.4.Hubbe_Gardner_Shen
5. Erdal, N., Hakkarainen, M.: Degradation of cellulose derivatives in laboratory, man-made, and natural environments. Biomacromol **23**(7), 2713–2729 (2022). https://doi.org/10.1021/acs.biomac.2c00336
6. Billingham, N., Knight, B., Garside, P., Mills, D.: Polymers and their degradation. Conservation of Books, Routledge, pp. 385–416 (2023). ISBN 978-0-36775-490-7
7. Bourdichon, F., Betts, R., Dufour, C., Fanning, S., Farber, J., McClure, P., et al.: Processing environment monitoring in low moisture food production facilities: are we looking for the right microorganisms? Int. J. Food Microbiol. **356**, 109351 (2021). https://doi.org/10.1016/j.ijfoodmicro.2021.109351
8. Wennerström, H., Oliveberg, M.: On the osmotic pressure of cells. QRB Disc. **3**, 12 (2022). https://doi.org/10.1017/qrd.2022.3
9. Francolini, I., Galantini, L., Rea, F., Di Cosimo, C., Di Cosimo, P.: Polymeric wet-strength agents in the paper industry: an overview of mechanisms and current challenges. Int. J. Mol. Sci. **24**(11), 9268 (2023). https://doi.org/10.3390/ijms24119268
10. Gupta, V., Biswas, D., Roy, S.: A comprehensive review of biodegradable polymer-based films and coatings and their food packaging applications. Materials **15**(17), 5899 (2022). https://doi.org/10.3390/ma15175899
11. Asgher, M., Qamar, S., Bilal, M., Iqbal, H.M.N.: Bio-based active food packaging materials: sustainable alternative to conventional petrochemical-based packaging materials. Food Res. Int. **137**, 109625 (2020). https://doi.org/10.1016/j.foodres.2020.109625
12. Tyagi, P., Salem, K., Hubbe, M., Pal, L.: Advances in barrier coatings and film technologies for achieving sustainable packaging of food products – a review. Trends Food Sci. Technol. **115**, 461–485 (2021). https://doi.org/10.1016/j.tifs.2021.06.036
13. Liu, B., Zhang, J., Guo, H.: Research progress of polyvinyl alcohol water-resistant film materials. Membranes **12**(3), 347 (2022). https://doi.org/10.3390/membranes12030347
14. Adibi, A., Trinh, B., Mekonnen, T.: Recent progress in sustainable barrier paper coating for food packaging applications. Prog. Org. Coat. **181**, 107566 (2023). https://doi.org/10.1016/j.porgcoat.2023.107566
15. Rungruangkitkrai, N., Phromphen, P., Chartvivatpornchai, N., Srisa, A., Laorenza, Y., Wongphan, P.: Water repellent coating in textile, paper and bioplastic polymers: a comprehensive review. Polymers **16**(19), 2790 (2024). https://doi.org/10.3390/polym16192790
16. Jongprateep, O., Mani-lata, C., Sakunrak, Y., Audcharuk, K., Narapong, T., Janbooranapinij, K., et al.: Titanium dioxide and fluoropolymer-based coating for smart fabrics with antimicrobial and water-repellent properties. RSC Adv. **12**(1), 588–594 (2022). https://doi.org/10.1039/D1RA05634D
17. Jacobs, S., Kosson, D., Brown, K., Cutshaw, A., Conrad, J., Feraldi, R., et al. Assessment of fluoropolymer production and use with analysis of alternative replacement materials. United States (2024). https://doi.org/10.2172/2370520
18. Xie, Y., Hill, C.A.S., Xiao, Z., Militz, H., Mai, C.: Silane coupling agents used for natural fiber/polymer composites: a review. Compos. A Appl. Sci. Manuf. **41**, 806–819 (2010). https://doi.org/10.1016/j.compositesa.2010.03.005
19. Kakar, P., Singh, A., Sheikh, J.: Flame retardant finishing of textiles - a comprehensive review. Indian J. Fibre Text. Res. **48**(4), 475–494 (2023). https://doi.org/10.56042/ijftr.v48i4.7662
20. Cappelletto, E., Callone, E., Campostrini, R., Girardi, F., Maggini, S., della Volpe, C., et al.: Hydrophobic siloxane paper coatings: the effect of increasing methyl substitution. J. Sol-Gel Sci. Technol. **62**(3), 441–452 (2012). https://doi.org/10.1007/s10971-012-2747-1

21. Taylor, R., Brenda, H.: Investigation of hydrophobic coatings on cellulose-fiber substrates with in-situ polymerization of silane/siloxane mixtures. Int. J. Adhes. Adhes. **86**, 13–21 (2018). https://doi.org/10.1016/j.ijadhadh.2018.07.008
22. Khosravani, A., Asadollahzadeh, M., Rahmaninia, M., Bahramifar, N., Azadfallah, M.: The effect of external and internal application of organosilicon compounds on the hydrophobicity of recycled OCC paper. BioResources **11**, 8257–2568 (2016). https://doi.org/10.15376/biores.11.4.8257-8268
23. Profili, J., Babaei, S., Al Rashidi, M., Dorris, A., Asadollahi, S., Sarkissian, A., et al.: Plasma-deposited organosilicon hydrophobic coatings on cellulosic materials for wet packaging applications. Coatings **13**(5), 924 (2023). https://doi.org/10.3390/coatings13050924
24. Bainbridge, A.: Material-Based Treatments. Conservation of Books, London (2023). https://doi.org/10.4324/9781003162674
25. Bhandari, V., Jose, S., Badanayak, P., Sankaran, A., Anandan, V.: Antimicrobial finishing of metals, metal oxides, and metal composites on textiles: a systematic review. Ind. Eng. Chem. Res. **61**(1), 86–101 (2022). https://doi.org/10.1021/acs.iecr.1c04203
26. Ly, B., Thielemans, W., Dufresne, A., Chaussy, D., Belgacem, M.: Surface functionalization of cellulose fibres and their incorporation in renewable polymeric matrices. Compos. Sci. Technol. **68**(15), 3193–3201 (2008). https://doi.org/10.1016/j.compscitech.2008.07.018
27. Etale, A., Onyianta, A., Turner, S., Eichhorn, S.: Cellulose: a review of water interactions, applications in composites, and water treatment. Chem. Rev. **123**(5), 2016–2048 (2023). https://doi.org/10.1021/acs.chemrev.2c00477
28. Sharmila, B., Chandra, J., Sasi, S., Arundhathi, K.: Modification of cellulose. In: Thomas, S., Hosur, M., Pasquini, D., Jose Chirayil, C. (eds). Handbook of Biomass. Springer, Singapore (2024). https://doi.org/10.1007/978-981-99-6727-8_22
29. Abdelmouleh, M., Boufi, S., ben Salah, A., Belgacem, M.N., Gandini, A.: Interaction of silane coupling agents with cellulose. Langmuir **18**(8), 3203–3208 (2002). https://doi.org/10.1021/la011657g
30. Merezhko, N., Komakha, V., Komakha, O., Tkachuk, V., Rechun, O.: Protection of Paper Surface from Water Wetting by Two-Layer Siloxane (TEOS/PEHS) Coating. In: Ivanov, V., Trojanowska, J., Pavlenko, I., Rauch, E., Peraković, D. (eds.) Advances in Design, Simulation and Manufacturing V. DSMIE 2022. LNME, Springer, Cham (2022). https://doi.org/10.1007/978-3-031-06025-0_45
31. Soytürk, E., Kartal, S., Onses, M., Celik, N.: Preliminary evaluation of polydimethylsiloxane and hydrophobic silica nanoparticles to improve water repellency and boron leachability in wood. Eur. J. Wood Wood Prod. **81**(1), 89–98 (2023). https://doi.org/10.1007/s00107-022-01883-7
32. Zhang, S., Zhang, X., Shang, W.: Chemical investigation of potassium methyl siliconate as deacidification and strengthening agent for preservation of aged papers. Chin. J. Polym. Sci. **33**(12), 1672–1682 (2015). https://doi.org/10.1007/s10118-015-1713-0

Wear Behaviour and Machining Performance of PrimeCon Coating in Milling of Ti-6Al-4V

Naiara Poli Veneziani Sebbe[1,2](✉), Filipe Daniel Fernandes[1,3], Heide Heloise Bernardi[4], Jorge Tadao Matsushima[4], and Rita de Cássia Mendonça Sales-Contini[1,4]

[1] ISEP, School of Engineering, Polytechnic of Porto, 4249-015 Porto, Portugal
napvs@isep.ipp.pt
[2] FEUP - Faculty of Engineering, University of Porto, 4200-465 Porto, Portugal
[3] University of Coimbra, CEMMPRE - Centre for Mechanical Engineering Materials and Processes, 3030-788 Coimbra, Portugal
[4] Technological College of São José Dos Campos, Professor Jessen Vidal, Centro Paula Souza, São José Dos Campos, SP 12247-014, Brazil

Abstract. Ti-6Al-4V has an excellent combination of mechanical properties such as high corrosion resistance, low density, high specific strength and low modulus of elasticity, leading this alloy to be widely used in the aerospace industry. However, this alloy is considered a difficult-to-machine material. It was found that the use of thin films deposited on the surface of cutting tools improves the performance and profitability of the process, as well as the machined surface quality, and decreases the resulting wear. Thus, this work aimed to evaluate the wear behaviour of coated cutting tools in the milling of Ti-6Al-4V, varying the cutting speed, cutting length, and feed per tooth. The coating deposited on the cutting tools had a thickness of 1.6 μm, substoichiometric chemical composition, and compact columnar morphology. It was found to belong to the TiAlSiN system. The surface roughness of the machined part, tool wear, and wear mechanisms under the considered conditions were analyzed. It was found that the feed-per-tooth has a strong influence on the wear of the cutting tools and the surface roughness of the machined part. Cutting speed also influence the process, with the increase in this parameter, increasing the flank wear. Regarding the wear mechanisms identified, the wear is predominantly abrasive. Adhesive wear was also observed, mainly for higher feed-per-tooth values. In addition, the cutting edge breakage was observed. Therefore, the wear infringed to the uncoated tools was effectively greater than in the coated tools, at all cutting speeds tested.

Keywords: Titanium alloy · Coated tool · Wear mechanisms · PVD coatings · Wear resistance

1 Introduction

The machinability of titanium alloys has been a topic of great interest for industry and scientific research worldwide. These alloys are widely used in the aerospace, biomedical, automotive, and naval industries due to their excellent mechanical properties. The use

F. J. G. Silva et al. (Eds.): DSMIE 2025, LNME, pp. 43–53, 2025.
https://doi.org/10.1007/978-3-031-96413-8_5

of titanium alloys has increased exponentially, especially in the aerospace industry, due to their excellent corrosion resistance and their good strength/weight ratio [1]. However, these excellent properties cause high wear on the cutting tool, making them difficult-to-machine material. During the machining process, conventional tools wear out very quickly due to the low thermal conductivity of the alloys, causing high temperatures near the cutting edge and adhesion of the workpiece material to the cutting tool. The analysis and identification of suitable machining conditions are key to obtaining a good surface finish for the workpiece and an increase in tool lifespan [2, 3].

This type of wear is related to several factors such as tool geometry, tool base material, properties of the material to be milled, cutting conditions, and milling parameters [3]. Another factor is the chemical reactivity of the material to be machined with the tool material, which causes adhesion welding to the cutting tool. Excessive heat that accumulates in the cutting zone has been identified as one of the main threats to surface integrity [4]. The greater the hardness of the material, the greater the machining forces required to remove the material, increasing by cutting temperatures [5]. These adverse conditions deteriorate the tool more aggressively both mechanically, for example, by abrasion, and thermally, for example, by oxidation [6]. In addition, factors such as vibration during the process, high stresses, chemical reactions, as well as high temperatures in the cutting region contribute to reducing tool performance [7]. Factors such as cutting speed, depth of cut, feed per tooth, use or not of coolants, and tool materials contribute significantly to the result obtained [2, 8]. Thus, the choice of cutting tools is crucial to obtaining good surface quality, efficiency, and economy in the machining process. To improve the performance of the milling process, thin coatings have been deposited on cutting tools. The coating selection is made according to the operating conditions, the material to be machined, thermal stability, tribological performance, and the affinity of the coating with the base material to be machined [6]. Based on that, this work aimed to analyze and compare the wear behaviour and process performance of tools without coating and tools coated with the commercial coating "PrimeCon". This coating was selected due to the gap seen in the literature regarding their analysis in the Ti-6Al-4V milling process. The results obtained here are expected to be used as a basis for future work in terms of optimization and performance of the milling process of titanium alloys.

2 Literature Review

2.1 Importance of Titanium Alloys in Industry

Titanium alloys are produced by combining titanium with various alloying elements. This combination contributes to the fact that Ti alloys are lightweight, and have high mechanical strength, excellent toughness, good corrosion resistance, and elevated resistance to high temperatures. These alloys help to replace heavy materials, increasing safety standards and thus reducing the number of accidents and failures. Due to these excellent properties, Ti alloys are promising for the aerospace industry [9]. They are currently used in jet engine parts, aircraft bodies, fan blades, engine compartments, and fuel tanks, among others [10]. Due to the great expansion of the aerospace industry, the demand for this type of alloy has grown exponentially, intending to improve the machinability of these alloys. This growth is largely due to the increase in the construction of high-power and high-performance, low-weight engines for military aircraft [11]. In the aerospace industry, machining is one of the most important processes present in the manufacture of parts. The main objective is to obtain a product that meets the design specifications, so it is important to monitor the condition of the tool and control the cutting parameters and the machining conditions [12].

2.2 Ti-6Al-4V and Machining Process

Ti-6Al-4V is considered a difficult-to-machine material. As a result, cutting tools used in machining this alloy suffer premature wear at the cutting edge. Excessive heat that accumulates in the cutting zone has been identified as one of the main threats to surface integrity [4]. Despite all the Ti-6Al-4V excellent properties, it is very difficult to machine due to high hardness at high temperatures, compared to steels with similar properties; variations in cutting and axial components cause vibrations during the process [13]. Furthermore, the Ti-6Al-4V alloy is typically difficult to machine using cutting speeds above 30 m/min using tools made of high-speed steel and above 60 m/min using cemented carbide tools, resulting in very low productivity [14]. Cutting speed is the parameter that most affect the temperature reached in the cutting zone and consequently tool wear [15]. Chip generation and cutting-edge wear are the main failure modes in the milling of titanium alloys with carbide tools. These types of failure modes are the result of a combination of several factors, such as high thermomechanical and cyclic stresses, as well as adhesion and tearing of the tool faces of the workpiece material [16]. The intense sliding action between the chip that is generated and the cutting tool causes the temperature at the contact to increase. The hot surface in contact with the tool surface causes a rapid chemical reaction, which causes significant wear to the tool [17]. Another critical type of wear is abrasive wear caused by small fragments of material [14]. Because some components for the aeronautical industry have complex geometries, such as thin ribs, corners, etc., makes the machining process even more difficult [18]. Given these difficulties, cutting tools with surface properties are necessary. Good hardness and wear resistance, toughness and thermal stability are the main properties required [19].

2.3 Solutions Used to Improve Machining

The use of solid lubricants helps to solve, to a certain extent, the dangerous effects of conventional cutting fluids, but does not eliminate the problem completely [20]. As there is adhesion between the solid lubricant, the machined surface, and the machine tool, additional cleaning is required. The disadvantages of using cutting fluids and solid lubricants can be overcome by depositing a thin film over the cutting tools [21]. This type of coating acts as a thermal barrier, providing lubrication and wear resistance during the machining process. The benefits arising from the use of coated cutting tools can be enhanced by using lubrication methods that use minimal amounts of lubricant, such as the MQL (Minimum Quantity Lubrification) technique [20]. This method has been widely used mainly when machining hard-to-cut materials or adherent materials [22]. It is more environmentally friendly, because it provides the part with a better surface finish and because the cutting tool is subjected to lower cutting forces, thus increasing its lifespan [23]. The authors Settireni and Faga [19] studied the turning process of titanium alloys with three cutting speeds (50, 90 and 130 m/min) and found that the use of the MQL system effectively increases the useful life of the tool due to its good lubrication capacity. Over the years, several scientific studies have been carried out to increase the tool lifespan used in machining titanium alloys. Their wear was investigated following the selected cutting parameters and the chosen base material [14]. The cemente carbide (WC/Co) continues to be the first choice in turning and milling of titanium alloys [16]. Thus, the high difficulty in machining titanium alloys and the constant search for more efficient solutions have contributed to an increase in the use of coatings to protect the surface of cutting tools. The coatings applied to cutting tools are generally nitrides, carbides, borides, transition metal oxides and carbonitrides [6]. Coatings without Ti in their composition can make the milling process of titanium alloys more efficient. The use of coatings based on ZrN or (Zr,Al)N can be quite effective [24].

3 Research Methodology

3.1 Materials

The machined alloy was Ti-6Al-4V. This alloy represents about 60% of the total titanium production [25]. Table 1 shows the chemical composition of this alloy. The workpiece material was supplied in the form of a parallelepiped measuring 100 mm in length, 50 mm in width and 80 mm in height.

Table 1. Ti-6Al-4V chemical composition (*%wt*).

Ti	Al	V	Fe	O
90	6	4	Max. 0.25	Max. 0.2

The cutting tools used in the milling process were end mills made of tungsten carbide (WC-Co), grade 6110, with an average grain size of 0.3 μm. Both the tools coated with the commercial coating "PrimeCon" (rose gold color) and the uncoated tools were supplied by INOVATOOLS, S.A. (Leiria, Portugal). The tools had a cutting diameter of 6 mm, cutting length of 13 mm, total length of 57 mm, chamfer 45°; 0.20 mm, helix angle of 35°/38°, rake angle of 25.4° and 4 blades.

3.2 Methods

Milling Tests

The machining strategy was defined taking into account the geometry of the part. Linear roughing machining was performed in up-milling and down-milling. As for the milling parameters, these were initially chosen taking into account the parameters recommended by the manufacturer. However, as the tests progressed, it became necessary to adjust the cutting parameters. The radial cutting depth used was 4.6 mm (77% of the tool diameter) and was kept constant to mitigate wear-related phenomena that could arise. The axial cutting depth was 0.5 mm. During the tests, cutting fluid, an emulsion composed of 5% synthetic base oil and 95% water, was used. The main objective was to evaluate the behavior of coated and uncoated tools for different cutting parameters. For this purpose, the following parameters were varied: cutting length, cutting speed, and feed per tooth. For easier identification of the cutters, the following nomenclatures were used: *UCTxFyLz* and *CTxFyLz*. In this nomenclature, the acronym 'UCT' represents uncoated cutters, and the acronym 'CT' represents coated cutters. The letter *x* represents the cutting speed value and can assume the values 60 m/min, 70 m/min, or 80 m/min. Next, the letter *y* represents the feed value per tooth which can present the values 0.026 mm/tooth, 0.03 mm/tooth, 0.07 mm/tooth, or 0.1 mm/tooth. Finally, the letter *z* represents the cutting length used in each test, which can take the values of 5 m, 10 m, or 15 m. It should be noted that the experiments were established in accordance with the parameters indicated by the manufacturer.

Analysis of Machined Surface Roughness, Wear and Coating Thickness
The roughness of the machined surface was measured using a MAHR PERTHOMETER M2 roughness tester (Mahr, Gottingen, Germany), following the DIN EN ISO 4288/ASME [26]. After each machining test, four measurements of the surface roughness were performed in the longitudinal direction of the cut, and the values obtained were used to calculate the arithmetic average roughness (*Ra*). Thus, it was possible to evaluate the stability of the process, the performance of the cutting tool and its correlation with the resulting wear in each tested condition. After the machining tests, the cutting tools were observed under a scanning electron microscope to characterize the chemical composition and morphology of the coating, as well as to know, evaluate and quantify the wear of the cutting tools. All tools were analyzed using a Hitachi scanning electron microscope (SEM), model SU3800. In the SEM analysis, the magnification range was adjusted depending on the area to be measured, allowing a better analysis of the coating morphology. An electron beam acceleration voltage of 6 kV was used, and the Backscattered Electron Detector (BSE) was used.Wear was measured according to ISO 8688–2:1989 [27]. This standard recommends analyzing the presence of all wear mechanisms and adopting as life criterion the one that has the greatest influence. Taking into account the type of wear present in the tested tools, and localized flank wear, the VB3 was selected.

4 Results and Discussion

4.1 Coating Characterization

Figure 1 shows the morphology of the cross-section´s coating. This coating presents a compact columnar morphology, although some columns can be seen, mainly in the upper part where the top of each column is visible. Columnar morphology is characteristic of the PVD deposition process [28]. The average thickness obtained was 1.6 μm.

Table 2 shows the chemical composition of the coating in atomic percentage. The chemical composition analysis shows that the coating is within the TiAlSiN system. The nitrogen (N) percentage reveals that the coating has a sub-stoichiometric composition since the ratio of the sum of all metallic elements divided by the nitrogen percentage is less than 1. Furthermore, although the coating contains C, this element may be present in the composition of the coating due to its surface contamination. It should be noted that the Al concentration is very small and that this percentage contributes minimally to improving its mechanical properties.

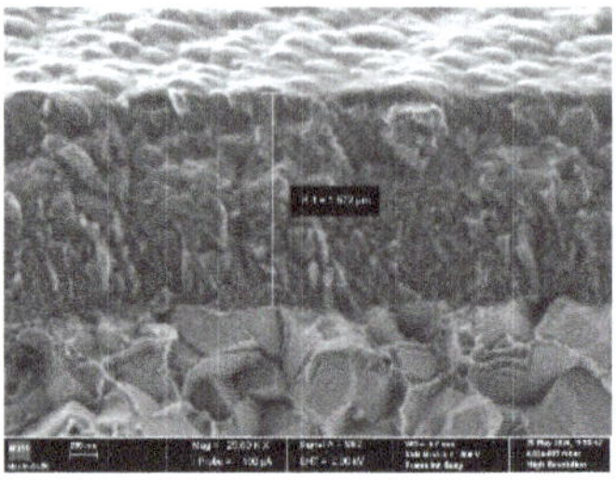

Fig. 1. Morphology and thickness of the coating.

Table 2. Coating chemical composition (%at)

C	N	Al	Si	Ti
4.1	54.6	0.5	8.9	33.9

4.2 Roughness Analysis

Figure 2 presents the results obtained from the machined part average surface roughness measured for all conditions, grouping them by coating or not, cutting speed, cutting length, and feed-per-tooth.

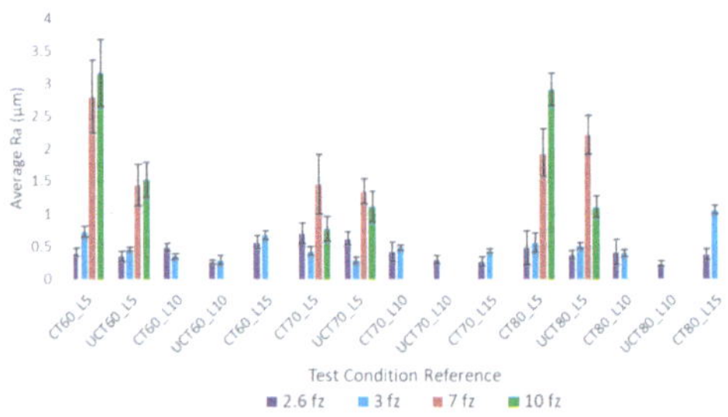

Fig. 2. Average surface roughness value (Ra) obtained for all conditions

For 60 m/min, it can be seen that *Ra* increased with the increase in feed per tooth, which was also obtained in the work by Oosthuizen et al. [29]. This trend is noticeable when the feed per tooth value increases by 0.07 mm/tooth and even greater when it is

0.1 mm/tooth. When analyzing the data for a cutting length of 5 m, it can be seen that the surface roughness of the uncoated tools is lower when compared to the coated cutting tools, for all feeds per tooth. This difference is greater for higher feed-per-tooth. Due to the high wear of the tools, it was only possible to analyze the cutting lengths of 10 m and 15 m for feeds per tooth of 0.026 mm/tooth and 0.03 mm/tooth, at all cutting speeds. For a cutting length of 10 m, very similar roughness values are observed for feeds per tooth of 0.026 mm/tooth and 0.03 mm/tooth. In the coated tools, the surface roughness was slightly higher for a feed per tooth of 0.03 mm/tooth.

For 70 m/min, the roughness values did not increase directly with the feed per tooth in these cases. For both coated and uncoated tools, the roughness was higher for a feed per tooth of 0.026 mm/tooth than for 0.03 mm/tooth and 0.07 mm/tooth than for 0.1 mm/tooth. This behavior change is attributed to the type of chip formed, wear mechanisms present, and cutting forces developed. For a 5 m cutting length, there is a clear increase in surface roughness when the feed per tooth is lower. For 10 m and 15 m, the roughness value is effectively higher as the feed per tooth increases, in all conditions tested since Ti-6Al-4V, as already mentioned, is a difficult-to-machine material.

Regarding 80 m/min of cutting speed, an effective increase in roughness can be observed with increasing feed per tooth. For 5 m, the increase in roughness is clear with increasing feed per tooth, while for uncoated cutters the roughness value is higher at 0.07 mm/tooth. This difference is due to the high wear caused by the tool. For 10 m, the roughness values were very similar, taking into account the feed per tooth, however, it can be seen that it is higher for coated tools than uncoated ones. At 15 m, the roughness was much higher for the highest feed per tooth value tested, this being the highest value obtained compared to the other cutting speeds.

4.3 Flank Wear Measurement (VB3) and Wear Mechanisms

Figure 3 shows the results obtained for the VB3 of the tools tested under all conditions, grouping them by coating or not, cutting speed, cutting length, and feed per tooth.

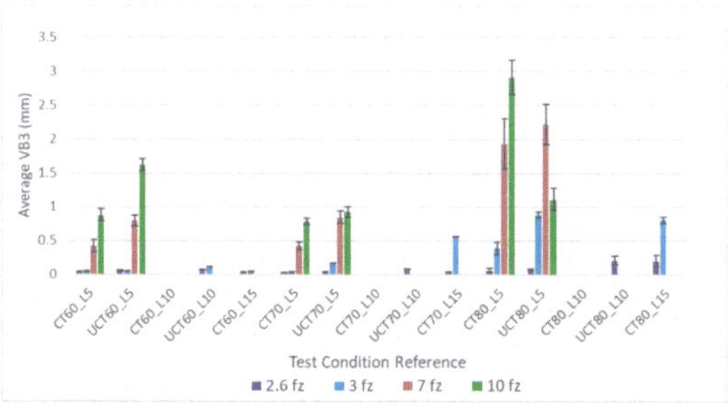

Fig. 3. VB3 for all tested conditions.

For 60 m/min, an increase in tool wear is observed with increasing feed per tooth. This trend was the same for both coated and uncoated tools, with greater wear being observed for uncoated tools. For the lowest feeds per tooth, 0.026 mm/tooth and 0.03 mm/tooth, very low wear is observed for all variations in cutting length, except for uncoated tools with a cutting length of 10 m. Furthermore, comparing the wear of uncoated tools with a cutting length of 10 m (UCT60_L10) to coated tools with a cutting length of 15 m (CT60_L15), it is observed that uncoated cutters present greater wear, thus showing the advantage of using coated tools.

At about 70 m/min, an increase in tool wear is observed with increasing feed per tooth. Comparing the results of coated and uncoated tools for 5 m, it can be seen that wear is greater on the uncoated tools (UCT_L5). Comparing these results with those obtained with a cutting speed of 60 m/min, the values are quite close, with a slight reduction in wear on the tools for a 5 m cutting length. Comparing the wear of uncoated tools with L = 10 m (UCT70_L10) with coated tools with L = 15 m (CT70_L15), it can be seen that wear, for the lowest feed value, is effectively greater for the uncoated tool with a cutting length of 10 m (UCT70_L10). However, a substantial increase in wear was observed on the coated tool for a cutting length of 15 m, with a feed rate of 0.03 mm/tooth. Furthermore, when comparing this result with that obtained with a cutting speed of 60 m/min, it was found that wear increased with speed. Therefore, the higher the cutting speed and the cutting length, the greater the wear caused to the tool. Regarding 80 m/min, an increase in tool wear is observed with the increase in feed per tooth, regardless of the cutting length value. For 5 m, greater wear is observed on the uncoated tool with 0.1 mm/tooth, compared to the coated tool. In general, with the increase in cutting speed to 80 m/min, an increase in wear is observed on the milling cutters, with the increase in feed per tooth, both on coated and uncoated tools, when compared to the lower speed conditions. The wear of the uncoated tool for 10 m (UCT80_L10) and 0.026 mm/tooth is practically the same as that obtained for the coated milling tool for 15 m (CT80_L15). However, with the increase in feed to 0.03 mm/tooth, the wear of the CT80_L15 milling tool increased severely, this being the highest value reached for these conditions. Thus, it is confirmed that effectively the higher the cutting speed and the cutting length, the greater the wear induced on the tool. Normally, increasing the cutting speed causes a decrease in wear, however, the feed per tooth is related to the chip generation mechanism, which directly influences the wear resistance of the cutting tools, and there may be an optimal value for the process to be optimized. The same was observed in the work of Li et al. [30], that is, an increase in flank wear with the increase in cutting speed, however, the authors used tools with polycrystalline diamond (PCD). In addition, in the work of Zhang et al. [31], when using carbide inserts, flank wear increased over time, and with the progression of the tests. Thus, the relationship between flank wear is verified regardless of the type of cutter used.

Predominant wear mechanisms were abrasive and adhesive on the tool substrate and the coatings. This is common in machining this kind of alloy [32]. With the increase in feed per tooth, there was an increase in tool wear, with a greater impact on uncoated tools. Liang et al. [33], observed the combined adhesion-diffusion-abrasion mechanisms in the milling of Ti6Al4V. In turn, Zheng et al. [34] found abrasion marks, material adhesion, built-up edges and micro-collapse blades in the milling of this alloy. In addition, the cutting edge breakage was observed in the coated tool for 0.1 mm/tooth and in the uncoated tools for 0.07 mm/tooth and 0.1 mm/tooth. For the condition with the highest cutting length value (L = 10 m), there was a slight adhesion of material on the uncoated cutter. Thus, in general, it was found that with the increase in the feed per tooth value, the tools presented greater wear, with an increase in abrasive wear and tool breakage, with the change in tool geometry. This was also observed by Dearnley and Grearson [35] when cutting titanium alloys, which had the geometry of the cutting tool modified as a result of the process. Furthermore, as in the work of Oosthuizen et al. [29], high chip generation was observed. This is related to the feed per tooth. In addition, it was found that the wear behavior of coated tools is better, i.e., qualitatively less wear generation was observed in coated tools, which contributes to the benefits of using coatings in the milling of Ti-6Al-4V. Figure 4 illustrates the resulting wear in the CT80F10L5 condition.

Fig. 4. Predominant wear mechanisms in CT80F10L5.

5 Conclusions

This work describes the performance of Ti-6Al-4V milling tools coated with the commercial coating "PrimeCon" and uncoated tools. The SEM analysis revealed that the coating deposited on the cutting tools had a thickness of 1.6 μm, a sub-stoichiometric chemical composition, belonging to the TiAlSiN system and a compact columnar morphology. Regarding the machining tests, it was observed that, mostly, the roughness value increased with the increase of the feed per tooth, both in coated and uncoated tools. The roughness values, for most of the feeds per tooth and cutting speeds tested, are lower than those obtained for the coated tools. On the other hand, because the uncoated tools presented broken teeth at low feeds per tooth values, showed that they have low fracture toughness and low wear resistance for the required conditions. The wear infringed to the uncoated tools was effectively greater than in the coated tools, at all cutting speeds tested. The wear mechanisms observed in the coated and uncoated tools were quite similar, however in greater quantity in the uncoated tools. The most predominant failure mechanisms were abrasive wear and adhesive wear. Furthermore, fracture of the cutting edge with geometric alteration was also observed.Thus, based on flank wear, the optimum condition for both coated and uncoated cutters was 70F2.6L5 The results obtained can be used as a basis for future work in which an analysis of vibrations occurring in the process can be performed, as well as, to verify its temperature.

Acknowledgment. The authors thank ISEP, INEGI, and FEUP for their institutional support. The authors also acknowledge Ricardo Alexandre from INOVATOOLS company, due to his availability to provide the tools used in this work free of charge, as well as Victor Moreira, due to his collaboration in the machining tests.

References

1. Teixeira, O., Silva, F.J.G., Ferreira, L.P., Atzeni, E.: A review of heat treatments on improving the quality and residual stresses of the Ti–6Al–4V parts produced by additive manufacturing. Metals **10**(8), 1006 (2020). https://doi.org/10.3390/met10081006
2. Hourmand, M., Sarhan, A., Sayuti, M., Hamdi, M.: A comprehensive review on machining of titanium alloys. Arabian J. Sci. Eng. **46**, 7087–7123 (2021). https://doi.org/10.1007/s13369-021-05420-1
3. Sousa, V.F.C., Silva, F.J.G.: Recent advances on coated milling tool technology—a comprehensive review. Coatings **10**(3), 235 (2020). https://doi.org/10.3390/coatings10030235
4. Carou, D., Rubio, E.M., Agustina, B., Marín, M.M.: Experimental study for the effective and sustainable repair and maintenance of bars made of Ti-6Al-4V alloy. Application to the aeronautic industry. J. Clean. Prod. **164**, 465–475 (2017). https://doi.org/10.1016/j.jclepro.2017.06.095
5. Sousa, V.F.C., et al.: Cutting forces assessment in CNC machining processes: a critical review. Sensors **20**(16), 4536 (2020). https://doi.org/10.3390/s20164536
6. Çalışkan, H., Kurbanoğlu, C., Panjan, P., Čekada, M., Kramar, D.: Wear behavior and cutting performance of nanostructured hard coatings on cemented carbide cutting tools in hard milling. Tribol. Int. **62**, 215–222 (2013). https://doi.org/10.1016/j.triboint.2013.02.035
7. Batista, M., Morales, A., Gómez-Parra, A., Salguero, J., Puerta, F.J., Marcos, M.: 3D-FEM based methodology for analysing contour milling. Procedia Eng. **132**, 1136–1143 (2015). https://doi.org/10.1016/j.proeng.2015.12.606
8. Martinho, R.P., Silva, F.J.G., Baptista, A.M.P.: Wear behaviour of uncoated and diamond coated Si_3N_4 tools under severe turning conditions. Wear **263**(7–12), 1417–1422. https://doi.org/10.1016/j.wear.2007.01.048
9. Fernández-Vidal, S.R., Mayuet, P., Rivero, A., Salguero, J., del Sol, I., Marcos, M:Analysis of the effects of tool wear on dry helical milling of Ti6Al4V alloy. Procedia Eng. **132**, 593–599 (2015). https://doi.org/10.1016/j.proeng.2015.12.536
10. Pushp P., Dasharath, S. Arati, C: Classification and applications of titanium and its alloys. Mater. Today Proc. **54**(2), 537–542 (2022). https://doi.org/10.1016/j.matpr.2022.01.008
11. Mruthunjaya, M., Yogesha, K.B.: A review on conventional and thermal assisted machining of titanium based alloy. Mater. Today Proc. **46**(17), 8466–8472 (2021). https://doi.org/10.1016/j.matpr.2021.03.490
12. Festas, A., Ramos, A., Davim, J.P.: Machining of titanium alloys for medical applications – a review. Proc. Inst. Mech. Eng. Part B: J. Eng. Manuf. **236**(4), (2021). https://doi.org/10.1177/09544054211028531
13. de Lacalle, L.N.L., Perez, J., Llorente, J.I., Sanchez, J.A.: Advanced cutting conditions for the milling of aeronautical alloys. J. Mater. Proc. Tech. **100**(1–3), (1998). https://doi.org/10.1016/S0924-0136(99)00372-6
14. Su, H., Liu, P., Fu, Y., Xu, J.: Tool life and surface integrity in high-speed milling of titanium. Chin. J. Aeronaut. **25**(5), 784–790 (2021). https://doi.org/10.1016/S1000-9361(11)60445-7
15. Martinho, R.P., Silva, F.J.G., Baptista, A.M.P.: Cutting forces and wear analysis of Si3N4 diamond coated tools in high speed machining. Vacuum **82**(12), 1415–1420 (2008). https://doi.org/10.1016/j.vacuum.2008.03.065
16. Jawaid, A., Sharif, S., Koksal, S.: Evaluation of wear mechanisms of coated carbide tools when face milling titanium alloy. J. Mater. Process. Technol. **99**(1–3), 266–274 (2000). https://doi.org/10.1016/S0924-0136(99)00438-0
17. Silva, F.J.G., Martinho, R.P., Martins, C., Lopes, H., Gouveia, R.M.: Machining GX2CrNiMoN26-7-4 DSS alloy: wear analysis of TiAlN and TiCN/Al_2O_3/TiN coated carbide tools behavior in rough end milling operations. Coatings **9**(6), 392 (2019). https://doi.org/10.3390/coatings9060392

18. Zhao, W., Wang, S., Han, Z., He, N.: Cutting performance evaluation of end mills for titanium aircraft componentes. Procedia CIRP **35**, 1–7 (2015). https://doi.org/10.1016/j.procir.2015.08.051
19. Settineri, L., Faga, M.G: Nanostructured cutting tools coatings for machining titanium. Mach. Sci. Tech. **12**(2), (2008). https://doi.org/10.1080/10910340802067239
20. Pedroso, A.F.V., et al.: Cooling and lubricating strategies for INCONEL® alloys machining: a comprehensive review on recent advances. J. Tribol. **1476**(6), (2024). https://doi.org/10.1115/1.4066955
21. Baptista, A., Silva, F., Porteiro, J., Míguez, J., Pinto, G.: Sputtering physical vapour deposition (PVD) coatings: a critical review on process improvement and market trend demands. Coatings **8**(11), 402 (2018). https://doi.org/10.3390/coatings8110402
22. Singh, R.: Minimum quantity lubrication turning of hard to cut materials – a review. Mater. Today: Proc. **37**(2), 3601–3605 (2021). https://doi.org/10.1016/j.matpr.2020.09.769
23. Carou, D., Rubio, E.M., Davim, J.P.: A note on the use of the minimum quantity lubrication (MQL) system in turning. Ind. Lubr. Tribol. **67**(3), 256–261 (2015). https://doi.org/10.1108/ILT-07-2014-0070
24. Vereschaka, A., Gurin, V., Oganyan, M., Oganyan, G., Bublikov, J., Shein, A.: Increase in tool life for end milling titanium alloys using tools with multilayer composite nanostructured modified coatings. Procedia CIRP **81**, 1412–1416 (2019). https://doi.org/10.1016/j.procir.2019.04.173
25. Liu, Z., An, Q., Xu, J., Chen, M., Han, S.: Wear performance of (nc-AlTiN)/(a-Si_3N_4) coating and (nc-AlCrN)/(a-Si_3N_4) coating in high-speed machining of titanium alloys under dry and minimum quantity lubrication (MQL) conditions. Wear **305**(1–2), 249–259 (2013). https://doi.org/10.1016/j.wear.2013.02.001
26. ISO 4288:1996 – Geometrical Product Specifications (GPS) – Surface texture: Profile method – Rules and procedures for the assessment of surface texture - International Organization for Standardization: Geneva, Switzerland (1996)
27. ISO 8688-2:1986 - Tool Life Testing in Milling – Part 2: End Milling; International Organization for Standardization: Geneva, Switzerland (1986)
28. Mauer, G., Vaßen, R.: Coatings with columnar microstructures for thermal barrier applications. Adv. Eng. Mater. **22**(6), 1900988 (2020). https://doi.org/10.1002/adem.201900988
29. Oosthuizen, G.A, Conradie, P.J.T., Dimitrov, D.M., Nunco, K.:The effect of cutting parameters on surface integrity in milling Ti6Al4V. South Afr. J. Ind. Eng. **27**(4)
30. Li, A., Zhao, J., Wang, D., Zhao, J., Dong, Y.: Failure mechanisms of a PCD tool in high-speed face milling of Ti–6Al–4V alloy. Int. J. Adv. Manuf. Tech. **67**, 1959–1966 (2013). https://doi.org/10.1007/s00170-012-4622-1
31. Zhang, S., Li, J.: Tool wear criterion, tool life, and surface roughness during high-speed end milling Ti-6Al-4V alloy. J. Zhejiang Univ.-Sci. A **11**, 587–595 (2010). https://doi.org/10.1631/jzus.A0900776
32. Jawaid, A., Sharif, S., Koksal, S.: Evaluation of wear mechanisms of coated carbide tools when face milling titanium alloy. J. Mater. Process. Technol. **99**, 266–274 (2000). https://doi.org/10.1016/S0924-0136(99)00438-0
33. Liang, X., Liu, Z.: Tool wear behaviors and corresponding machined surface topography during high-speed machining of Ti-6Al-4V with fine grain tools. Tribol. Int. **121**, 321–332 (2018). https://doi.org/10.1016/j.triboint.2018.01.057
34. Zheng, T., Song, Q., Du, Y., Liu, Z.: Development of tool wear standards and wear mechanism for micro milling Ti-6Al-4V Alloy. Metals **12**(5), (2022). https://doi.org/10.3390/met12050726
35. Dearnley, P.A., Grearson, A.N.: Evaluation of principal wear mechanisms of cemented carbide and ceramics used for machining titanium alloy. IMI 318, Mater. Sci. Technol. **2**, 47–58 (1986). https://doi.org/10.1179/026708386790123611

Composite and Polymer Materials

The Effect of Repeated Processing of the Material on the Performance Properties of the Polymer Composite

Jozef Dobránsky(✉), Peter Michalík, Marek Kočiško, and Martin Pollák

Technical University of Košice, 31, Štúrova Street, Prešov 08001, Slovak Republic
jozef.dobransky@tuke.sk

Abstract. The issue of recycling polymer materials is especially important for more expensive polymer materials, but the cheaper so-called commodity plastics should not be underestimated either. The aim of the presented article is the monitoring of the repeated processing of polymer composite material for the utility properties of the material. A polymer composite with a polymer matrix polypropylene (PP), filled with particulate filler talc, was chosen for the analysis. Rheological and mechanical properties were chosen as useful properties. The melt flow volume index was chosen from the rheological properties. Tensile properties and impact resistance were selected from the mechanical properties using the Charpy method. Based on the measured rheological properties, we can conclude that repeated processing of the material slightly affects its fluidity. Based on the measured results of the mechanical properties, we can conclude that repeated processing of the material affects these properties.

Keywords: polymer · composite · properties · polypropylene · talc

1 Introduction

The growing interest in the preservation of our environment stimulates solutions for developing fewer impacting materials. Thus, the development of composites from renewable resources and recyclable end-of-life resources seems an interesting alternative [1–3].

The authors Henshaw et al. [4] and Graupner et al. [5] present an overview of the main problems related to the disposal of polymer composite materials after their use. They compare composites with unreinforced plastics in terms of production volume and available reprocessing technologies. Not surprisingly, composites make up a very small portion of plastic waste. However, in high-volume applications such as automobiles, there are reasons to be concerned about recycling composites.

In their study, Kuram et al. [6] and Sarrionandia et al. [7] investigated the recycling of glass fiber reinforced composites. From the experimental results, it was concluded that the recycled composites reinforced with glass fibers and talc filler showed better mechanical properties compared to the original binary blends, and the fifth recycled

F. J. G. Silva et al. (Eds.): DSMIE 2025, LNME, pp. 57–66, 2025.
https://doi.org/10.1007/978-3-031-96413-8_6

composite could be instead of the original material. It was also found that the recycled composite material was a suitable material for the recycling process and this polymer blend did not suffer any significant damage [8].

2 Literature Review

Mechanical recycling is a viable option for reducing plastic waste, especially for post-consumer waste and scraps from production processes. Biodegradable polymers have gained attention as alternatives to conventional ones, and their potential for reprocessing using standard techniques before composting needs to be explored [9–13].

In their studies, other authors devoted themselves to experimental investigation and modeled the effect of reworking on the quasi-static uniaxial tensile behavior of two commercial polypropylene (PP)-based composites. The studied materials mainly consist of unfilled impact-resistant PP and talc-filled impact-resistant PP. They subjected these PP composites to repeated processing cycles, including a grinding step and an extrusion step, to simulate laboratory-scale recycling. Repeated rework led to thermomechanical degradation by chain scission mechanisms. The tensile behavior of the two materials showed a continuous decrease in modulus and strain at failure with increasing amount of rework [14, 15].

A comprehensive survey of advances in polymer composites is currently needed, focusing on their performance, manufacturing processes, recycling techniques and sustainable practices. With the growing demand for high-performance materials in various industries, including aerospace, automotive and construction, polymer composites have emerged as a key solution offering unmatched strength, durability and lightweight properties. The research explores the fundamental characteristics of polymer composite materials and blends, emphasizing the synergy between different matrices and reinforcements that lead to improved material properties [16–20].

3 Research Methodology

3.1 Material Selection – Taboren PH 63 T 20

Table 1 shows the basic material properties of Taboren PH 63 T 20.

The material with the trade name Taboren PH 63 T 20 is a homopolymer polypropylene (PP) reinforced with 20% by weight of the mineral filler talc, which is also UV stabilized. The material is used to produce products that find application in the automotive industry in the production of various components in the interior and exterior of the car [21, 22].

3.2 Preparation of Tested Specimens

The procedure for preparing recycled material could be summarized in the following steps:

Production of the first set of test specimens from virgin material. In this step, 200 working strokes of the injection molding machine are produced, which represents 400 test specimens, since the test specimens are produced in a double injection mold.

Table 1. Material properties of Taboren PH 63 T 20.

Properties	Unit	Value
Density (23 °C)	g/cm3	1.04
Melt-Mass Flow Rate (2,16 kg / 230 °C)	g/10 min	7.0
Tensile Strength	MPa	33
Tensile Strain at Break	%	5
Flexural Modulus (23 °C)	MPa	2800
Charpy Unnotched Impact Strength (23 °C)	kJ/m2	2.4
Charpy Notched Impact Strength (23 °C)	kJ/m2	32

From the batch prepared in this way, 40 specimens are taken, which will be used to measure selected properties of plastic moldings. Other test specimens, including those with inlet systems, are crushed on a knife mill and recycled material is prepared. By this grinding, a second batch of material is prepared, which will be used for the preparation of test specimens, which represent the first stage of repeated recycling of the material. This procedure is repeated until the seventh stage of repeated material recycling. The preparation of test specimens took place under pre-planned conditions. The process of injection of the test specimens took place on a standard injection machine DEMAG Systec 100–310 with a universal two-plate injection mold according to the technological parameters listed in the material sheets.

After injection of test specimens from individual batches, these test specimens are marked so that they are not confused during testing. The marking of test specimens consists of a letter that represents the filler in the composite (T – talc) and a number that represents the degree of repeated recycling of the material. The principle of labeling test samples is shown in Table 2.

Table 2. Designation of material batches.

Batch	Description of the batch
T0	Virgin material (without previous processing)
T1	Material after the first transition
T2	Material after the second transition
T3	Material after the third transition
T4	Material after the fourth transition
T5	Material after the fifth transition
T6	Material after the sixth transition
T7	Material after the seventh transition

3.3 Establishing Rheological Properties

The melt flow index (MFI) is a standardized technological test of thermoplastic materials that serves to evaluate the flow properties. The Melt Volume Flow Rate (MVR) was chosen for testing. The MVR was determined by a Thermo Haake Meltflow MT rheometer (Fig. 1) according to the ISO 1133 standard [23].

Fig. 1. Rheometer MeltFlow Haake MT.

3.4 Establishing Tensile Properties

The test is conducted under the specified pretreatment conditions, ambient and speed parameters of the test bodies defined by ISO 527–1/2 [24]. The test bodies were stretched in the direction of their main longitudinal axis until disrupted, at a constant speed of 5 mm/min. During the test, the values of the applied load and the increase in the initially measured length of the test piece were recorded. 10 test bodies from each batch were tested on the Tiratest 2300 tearing machine (Fig. 2).

The test body was placed and fastened to the clamps of tensile machine so that the longitudinal axis of the body was identical to the axis of the machine and so that release of the test piece would be prevented. The module of tensile elasticity was determined at a reduced test speed 1 mm/min., measured by Epsilon – model 3542-010M-025-ST strain gauge.

3.5 Establishing Impact Properties

Determination of unnotched impact strength by the Charpy method was carried out under the conditions specified in the ISO 179–1 standard [25]. Test will be executed in the specimens conditioned at room temperature and at low temperature. Before the test, the span of the instrument brackets was set, and the instrument was calibrated. Afterwards, a hammer was let down loosely to determine the friction losses. At the same time, it was checked whether the Ceast Resil 5.5 (Fig. 3) instrument was able to perform

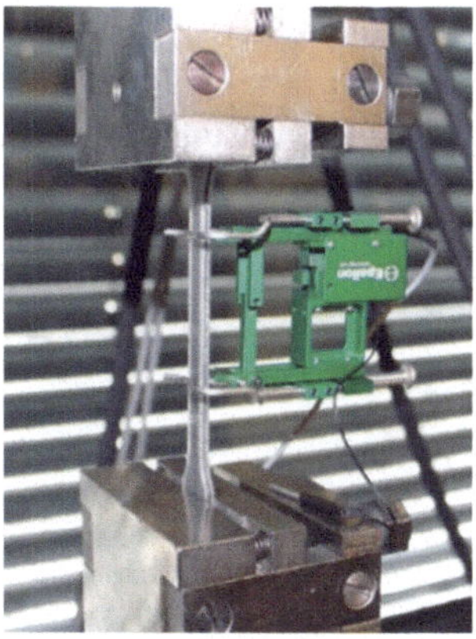

Fig. 2. Placement the strain gauge on the test sample on the Tiratest 2300.

the test at the prescribed impact velocity and whether it was in the correct range of the energy absorbed. For the purposes of experimental measurement, an impact hammer with a nominal energy of 2 J was used. Half of the test specimens intended to determine the unnotched impact strength at low temperatures were conditioned in the ProfiMaster PMU 0450 laboratory freezer at −30 °C for 24 h. A thorough type (C) puncture has been recorded in all test specimens.

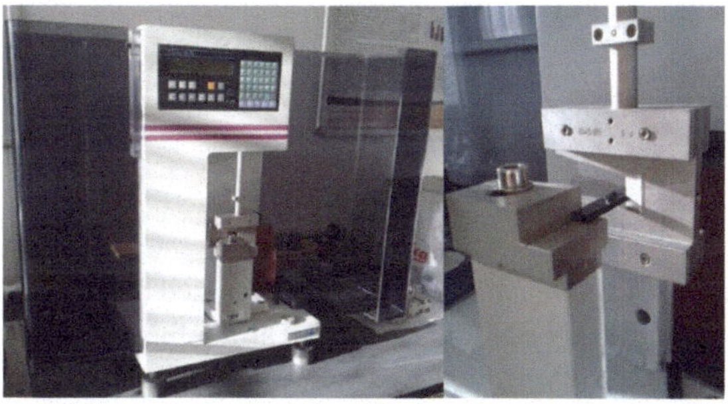

Fig. 3. Test device CEAST Resil 5.5 with sample location.

4 Results and Discussion

4.1 Evaluation of the Melt-Volume Flow Rate (MVR)

The graphical dependence of MVR on the degree of repeated recycling of the material is shown in Fig. 4. As we can see from the graphical dependence, the gradual repeated recycling of the material leads to an increase in the value of the melt flow index.

From the experimental measurement of MVR, which is closely related to the molecular weight, this method is clearly suitable for the detection of recycled material and not only in relation to the measured results. From a practical point of view, it is also important, because equipment for determining the melt flow index is generally available to all larger plastics processors, or at least they ensure this measurement externally regarding the requirements of their customers.

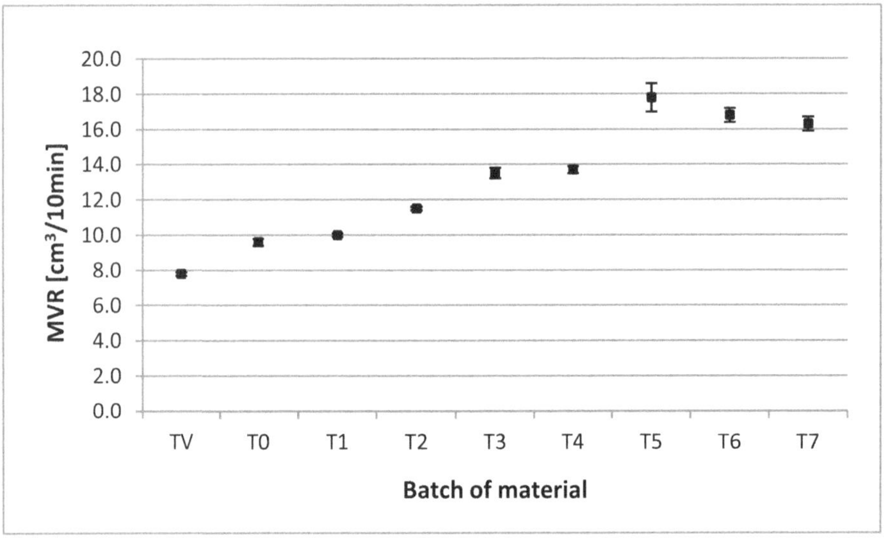

Fig. 4. The effect of the number of material transitions (batch) on the Melt-Volume Flow Rate (MVR).

4.2 Evaluation of the Tensile Properties

The determination and evaluation of the properties obtained by the tensile test took place according to the standard STN EN ISO 527–1/2. During the test, the dependence of the tensile stress on the relative elongation was recorded.

The ultimate tensile stress at break σ_b is the same as the tensile strength σ_m, which characterizes the maximum value of the tensile strength, since it was a measurement of the tensile properties of composite materials. The effect of the number of material transitions on the tensile strength is shown on Fig. 5.

Figure 6 show effect of the number of material transitions on the relative elongation. From the course of the tensile properties of the material PP T 20 and considering the dispersion of the measured values expressed by the standard deviation, it can be concluded that repeated recycling does not significantly affect the values of the tensile properties. The tensile strength value decreased from 35.2 MPa (T0) to 34.0 MPa (T7), which represents a decrease of only 3%. The ductility of the material, or the relative elongation decreased from the value of 6.4% (T0) to the value of 5.9% (T7).

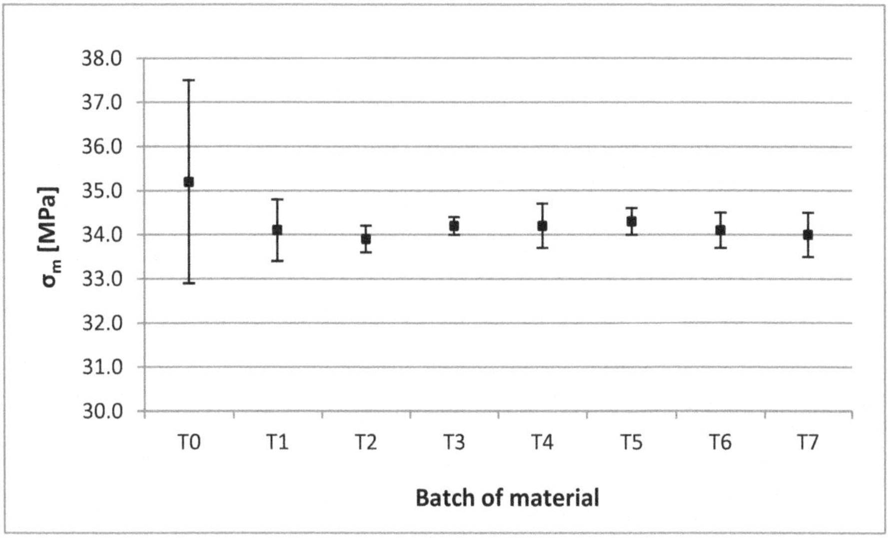

Fig. 5. The effect of the number of material transitions (batch) on the tensile strength (σ_m).

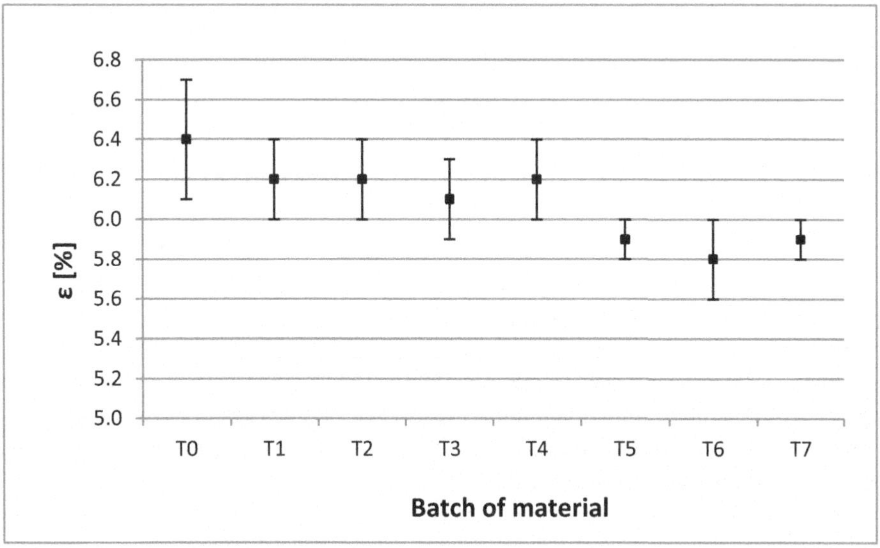

Fig. 6. The effect of the number of material transitions (batch) on the elongation (ε).

4.3 Evaluation of the Impact Properties

Impact resistance using the Charpy method was determined on test specimens in accordance with the international standard STN EN ISO 179–1. Impact resistance was measured similarly to tensile properties in two stages. First, the measurement took place at a laboratory temperature of 23 °C. The second part of the samples were conditioned at a temperature of -30 °C for 24 h in a laboratory freezer. The samples exposed in this way were subsequently tested.

The graphical dependence of impact resistance on the number of material transitions of PP T 20 material measured at different temperatures is shown in Fig. 7. As can be seen in the graph, the impact resistance value varied minimally, fluctuating around the value of 12.5 kJ/m^2 for all batches of material.

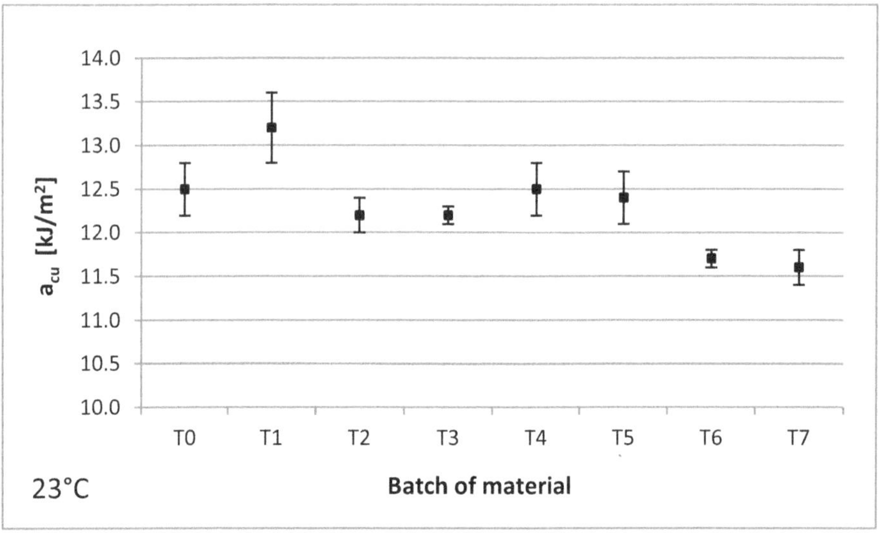

Fig. 7. The effect of the number of material transitions (batch) on the Charpy unnotched impact strength (a_{cu}).

As can be seen in the Fig. 7, during the impact resistance test, at a temperature of – 30 °C, all batches of PP T20 material were completely punctured by the test body.

The impact resistance value for bodies exposed to low temperature showed an identical fluctuating character during the measurement as it was for bodies that were not exposed to temperature. Individual impact resistance values measured on exposed bodies for individual batches of material decreased on average by approximately 3 kJ/m^2 compared to bodies that were not exposed.

5 Conclusions

Based on the methodology performed and the results achieved, it is possible to state that the reuse of recycled material is appropriate to a certain extent. On the other hand, it is possible to detect the presence of recycled material in products made of polymer materials with an appropriately chosen methodology.

- Based on the measured rheological properties, we can conclude that repeated processing of the material slightly affects its fluidity.

- Based on the measured results of the mechanical properties, we can conclude that repeated processing of the material does not affect these properties.

The main novelty and advantages of the obtained results lie in the introduction of a methodology for investigating the impact of recycled material on the performance properties of polymer materials.

Based on the proposed methodology, future research can also be focused on the area of reuse of recycled materials from PET bottles. In the near future, the European Commission plans to reuse recycled material from PET bottles in a volume of 30% by weight. The proposed methodology for monitoring the utility properties of these materials will serve here.

Acknowledgment. This research has been elaborated in the framework of the projects VEGA No. 1/0488/23 and KEGA No. 002TUKE-4/2023 were granted by the Ministry of Education, Science, Research and Sport of the Slovak Republic and this publication is the result of the Project implementation: Development of excellent research capacities in the field of additive technologies for the Industry of the 21st century, ITMS: 313011BWN5, supported by the Operational Program Integrated Infrastructure funded by the ERDF.

References

1. Krim, S., Boukerrou, A., Djidjelli, H., Beaugrand, J., Hammiche, D.: Reprocessing of composites based on polypropylene loaded with olive husk flour. Compos.-Mech. Comput. Appl. **9**(2), 163–187 (2018). https://doi.org/10.1615/CompMechComputApplIntJ.2018024680
2. Bhattacharjee, S., Bajwa, D.S.: Degradation in the mechanical and thermo-mechanical properties of natural fiber filled polymer composites due to recycling. Constr. Build. Mater. **172**, 1–9 (2018). https://doi.org/10.1016/j.conbuildmat.2018.03.010
3. Achukwu, EO., et al.: Effect of reprocessing on the service life properties of glass fiber-reinforced in-house polypropylene waste composites. Adv. Manuf.-Polym. Compos. Sci. **9**(1), (2023). https://doi.org/10.1080/20550340.2023.2195317
4. Adams, R.D., et al.: Recycling of reinforced plastics. Appl. Compos. Mater. **21**(1), 263–284 (2014). https://doi.org/10.1007/s10443-013-9380-1
5. Graupner, N., Albrecht, K., Ziegmann, G., Enzler, H., Mussig, J.: Influence of reprocessing on fibre length distribution, tensile strength and impact strength of injection moulded cellulose fibre-reinforced polylactide (PLA) composites. Express Polym Lett **10**(8), 647–663 (2016). https://doi.org/10.3144/expresspolymlett.2016.59
6. Kuram, E., Ozcelik, B., Yilmaz, F.: The influence of recycling number on the mechanical, chemical, thermal and rheological properties of poly(butylene terephthalate)/polycarbonate binary blend and glass-fibre-reinforced composite. J. Thermoplast. Compos. Mater. **29**(10), 1443–1457 (2016). https://doi.org/10.1177/0892705715569823
7. Biessey, P.: Chemical recycling of plastics. Chem. Ing. Tec. **95**(8), 1187 (2023). https://doi.org/10.1002/cite.202370802
8. Sergi, C., Tirillo, J., Iacovacci, C., Sarasini, F.: Influence of reprocessing cycles on the morphological, thermal and mechanical properties of flax/basalt hybrid polypropylene composites. Sustain. Mater. Technol. **36** (2023). https://doi.org/10.1016/j.susmat.2023.e00648
9. Bavasso, I., et al.: Recycling of a commercial biodegradable polymer blend: Influence of reprocessing cycles on rheological and thermo-mechanical properties. Polym. Test. **134** (2024). https://doi.org/10.1016/j.polymertesting.2024.108418

10. Uzosike, C.C., Yee, L.H., Padilla, R.V.: Small-scale mechanical recycling of solid thermoplastic wastes: a review of PET, PEs, and PP. Energies **16**(3), (2023). https://doi.org/10.3390/en16031406
11. Romani, A.; Caba, S.; Suriano, R.; Levi, M.: Recycling glass and carbon fibers for reusable components in the automotive sector through additive manufacturing. Appl. Sci.-Basel **13**(10), (2023). https://doi.org/10.3390/app13105848
12. Krehula, L.K.: Recycling of plastic waste. J. Chem. Chem. Eng. **64**(7–8), 417–420 (2015)
13. Behalek, L., Novak, J., Brdlik, P., Boruvka, M., Habr, J., Lenfeld, P.: Physical properties and non-isothermal crystallisation kinetics of primary mechanically recycled poly(l-lactic acid) and poly(3-hydroxybutyrate-co-3-hydroxyvalerate). Polymers **13**(19), 3396 (2021). https://doi.org/10.3390/polym13193396
14. Wang, K., et al.: Constitutive modeling of the tensile behavior of recycled polypropylene-based composites. Materials **12**(15), 2419 (2019). https://doi.org/10.3390/ma12152419
15. Kraft, A.: Catalytic recycling of PET plastic. Chem. Ind. **86**(5), 40 (2022)
16. Paraye, P., Sarviya, R.M.: Advances in polymer composites, manufacturing, recycling, and sustainable practices. Polym.-Plast. Technol. Mater. **63**(11), 1474–1497 (2024). https://doi.org/10.1080/25740881.2024.2339295
17. Ramesh, M., Rajeshkumar, L.N., Srinivasan, N., Kumar, D.V., Balaji, D.: Influence of filler material on properties of fiber-reinforced polymer composites: a review. E-Polymers **22**(1), 898–916 (2022). https://doi.org/10.1515/epoly-2022-0080
18. Maiti, S., Islam, M.R., Uddin, M.A., Afroj, S., Eichhorn, S.J., Karim, N.: Sustainable fiber-reinforced composites: a review. Adv. Sustain. Syst. **6**(11), 2200258 (2022). https://doi.org/10.1002/adsu.202200258
19. Ovsik, M., Fucikova, K., Manas, L., Stanek, M.: Influence of polymer flow on polypropylene morphology, micro-mechanical, and tribological properties of injected part. Lubricants **12**(6), 202 (2024). https://doi.org/10.3390/lubricants12060202
20. Baron, P., Kascak, J.: Comparison of selected materials intended for the manufacture of plastic holder for reverse parking sensor with the use of computer simulation tools. MM Sci. J. **2019**, 3463–3467 (2019). https://doi.org/10.17973/MMSJ.2019_12_2019017
21. Krasinskyi, V., Gajdos, I., Suberlyak, O., Antoniuk, V., Jachowicz, T.: Study of the structure and thermal characteristics of nanocomposites based on polyvinyl alcohol and intercalated montmorillonite. J. Thermoplast. Compos. Mater. **34**(12), 1580–1691 (2021). https://doi.org/10.1177/0892705719879199
22. Jachowicz, T., Gajdos, I.: Effect of natural ageing on some properties of oxybiodegrading agent-containing polypropylene products. Przem. Chem. **93**(11), 1983–1985 (2014)
23. ISO 1133-1:2022 Plastics — Determination of the melt mass-flow rate (MFR) and melt volume-flow rate (MVR) of thermoplastics. https://www.iso.org/standard/83905.html
24. ISO 527-1:2019 Plastics — Determination of tensile properties. https://www.iso.org/standard/75824.html
25. ISO 179:1:2023 Plastics – Determination of Charpy impact properties. https://www.iso.org/standard/84393.html

Mechanical Characterization of an Aluminum-Composite Hybrid Joint Bonded by Ultrasonic Welding

Rafael R. Lucas[1(✉)], Rogério Pinto Mota[1], Francisco J. G. da Silva[2], Edson C. Botelho[1], and Rita C. M. Sales-Contini[2,3]

[1] School of Engineering and Sciences, São Paulo State University (UNESP), Guaratinguetá, São Paulo 12516-410, Brazil
rr.lucas@unesp.br

[2] CCIDEM, ISEP, Polytechnic of Porto, Rua Dr. António Bernardino de Almeida, 431, 4249-015 Porto, Portugal

[3] Technological University of São José Dos Campos, Professor Jessen Vidal, Centro Paula Souza, São José Dos Campos, São Paulo 12247-014, Brazil

Abstract. Plasma Electrolytic Oxidation (PEO) is an advanced surface treatment process for light alloys such as Aluminum, Titanium, Magnesium, and others. It is similar to the anodizing process but uses environmentally friendly electrolytes, such as silicates, aluminates, and phosphates. Due to thermal and plasma phenomena, PEO enables the development of an oxide coating with enhanced physical and chemical properties compared to conventional anodizing. In this study, samples of plasma-anodized AA2024 aluminum alloy were welded to thermoplastic composite PEI/Glass Fiber samples using ultrasonic welding. The welded joint was mechanically characterized by a single lap shear test (based on ASTM D1002:19), and the fractures were evaluated according to ASTM D5573–99 (failure modes). Scanning Electron Microscopy (SEM) was performed on both treated and untreated samples to justify the obtained results. It was observed that the samples treated by the PEO process showed lower shear resistance, around 2.10% less, although regions with interlaminar failure in the composite were found. Additionally, fractures occurred within the oxide coating, indicating overgrowth of the oxide in some regions. SEM analysis revealed that the treated samples did not present pores as observed in the untreated samples, which contributed to fewer mechanical anchoring points between the polymer matrix and the aluminum sample, thereby reducing shear resistance.

Keywords: Ultrasonic welding · Hybrid structure · Composite · ASTM

1 Introduction

The integration of high-performance materials, such as carbon/glass fiber-reinforced polymers, in vehicle manufacturing is transforming the automotive and aerospace industries, promoting sustainability and energy efficiency. These materials, particularly thermoplastic composites, offer recyclability and reprocessability, significantly reducing

F. J. G. Silva et al. (Eds.): DSMIE 2025, LNME, pp. 67–75, 2025.
https://doi.org/10.1007/978-3-031-96413-8_7

material usage and environmental impact [1]. Furthermore, the application of such materials is driven by the fact that carbon fiber and glass fiber are lighter than traditional metals, increasing vehicle efficiency and reducing energy consumption during operation [2].

Although high-performance materials offer significant benefits, it is essential to overcome challenges related to production and lifecycle management to fully realize their potential in creating sustainable vehicles. The high initial cost of composite materials can limit their widespread adoption. Additionally, manufacturing processes can be energy-intensive, raising concerns about the environmental impact throughout the entire lifecycle of these materials [3, 4].

For its final application, this material is often joined with metals such as aluminum, titanium, or even high-alloy materials, requiring complex assemblies due to the differing chemical compositions and mechanical properties of the constituent materials. The most commonly used methods for this are adhesive bonding and mechanical fastening [1, 5].

2 Literature Review

Joining materials with significantly different properties is a major challenge, as it is necessary to ensure a surface with the appropriate level of roughness for effective mechanical interlocking and establish a certain degree of chemical bonding between the materials. To achieve this, surface treatments such as anodization for light alloys and techniques like CVD, sputtering, DBD, and others are commonly employed [6–8].

In alloys such as aluminum, the anodizing process is employed, which provides remarkable surface hardness and allows for better absorption of paints, thereby ensuring greater durability of structures [9, 10].

Although the anodizing process is highly efficient and economical, it presents environmental drawbacks, particularly in hard anodizing, which employs hexavalent chromium (Cr^{6+}), as well as the extensive effort required for pre- and post-treatment of parts and the effluents generated [10]. In this context, the PEO process stands out as it provides an oxide coating with excellent physical and chemical properties for the base material, making it suitable for aerospace applications and biomaterials. It also excels as a more environmentally efficient process, utilizing alkaline electrolytes based on silicates, aluminates, phosphates, etc. [11].

In this study, AA2024 aluminum alloy anodized by the PEO process was ultrasonically welded to a PEI/Glass fiber thermoplastic composite. Three samples in each condition (untreated and treated) were analyzed through lap shear mechanical testing, fractographic analysis, and scanning electron microscopy (SEM) of the surfaces.

3 Research Methodology

3.1 AA2024 and PEI/Glass Fiber Thermoplastic Composite

For this study, the alloy used was AA2024-O. Samples of this alloy were cut using a hydraulic shear, resulting in dimensions of 100 × 25 × 2.0 mm, based on ASTM D1002:2019 [12]. The chemical composition of the alloy, as provided by the supplier, is shown in Table 1.

Table 1. Chemical composition of AA2024 alloy, according to the supplier.

Al	Si	Fe	Cu	Mn	Cr	Zn	Ti
Balance	< 0.5%	< 0.5%	3.8–4.9%	0.3–0.9%	< 0.1%	< 0.25%	< 0.15%

This same alloy has been used in other studies [10], applying the oxyacetylene welding process to join the aluminum alloy to the PEI/Glass fiber thermoplastic composite. After cutting, the samples were sanded with SiC #100 sandpaper and cleaned using ultrasonic treatment in a solution of deionized water and neutral detergent. Subsequently, the samples underwent cleaning with isopropyl alcohol.

3.2 Treatment of AA2024 Alloy by PEO

For this study, the PEO treatment was employed to create a thin alumina (Al2O3) oxide layer on the alloy, as discussed in previous studies [9, 10]. The electrolyte used for this process was sodium tetraborate ($Na_2B_4O_7 \cdot 10H_2O$–7.0 g/L), with the addition of potassium hydroxide (KOH–1.0 g/L) to increase conductivity. The other anodization parameters used in this study are presented in Table 2.

Table 2. Chemical composition of AA2024 alloy, according to the supplier.

Time (s)	Duty Cycle (%)	Concentration (g/L)	Frequency (kHz)
324	15	7.0	25

3.3 Ultrasonic Welding and LSS Testing

The joining of the AA2024 aluminum alloy and the thermoplastic composite was performed using ultrasonic welding, utilizing an MCX CC21–01606 equipment from the manufacturer Mecasonic. The welding parameters are presented in Table 3.

Table 3. Chemical composition of AA2024 alloy, according to the supplier.

Start Pressure	Welding Pressure	Welding Time	Holding pressure	Holding Time	Ascent pressure
0.2 MPa	0.6 MPa	1.0 s	0.2 MPa	2.0 s	0.3 MPa

Three samples were welded under each condition (untreated and treated with PEO). After the welding process, the samples were tested according to the ASTM D1002:2019 standard [12], using a Shimadzu mechanical testing machine with a load module of 100 kN and a displacement speed of 1.5 mm/min.

3.4 Characterizations

The oxide coating produced by the PEO process was morphologically characterized using scanning electron microscopy (SEM) with an EVO LS15 microscope from Zeiss. The fracture resulting from the Lap Shear test on the aluminum and composite samples was analyzed using a digital microscope, model INSKAM-307, from Shenzhen Yipincheng Technology.

4 Results and Discussions

4.1 Shear Strenght (LSS)

The force vs. strain graphs of the hybrid joints are presented in Fig. 1. It is observed that the control samples (without PEO treatment) exhibited a maximum force of 1973.9 ± 263.9 N, while the samples with PEO treatment showed a maximum force of 1932.4 ± 692.4 N.

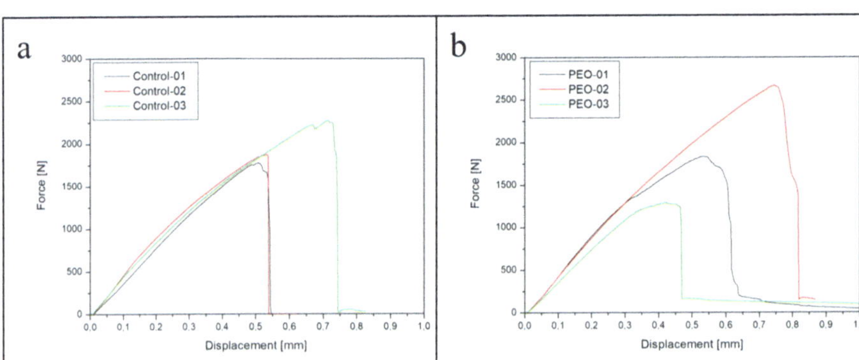

Fig. 1. Force vs. Strain graph of the hybrid joint, samples without treatment (a) and with PEO treatment (b)

The control samples exhibited similar graphs in terms of strain and final stress, with only one case significantly deviating. In contrast, the PEO-treated samples displayed considerable variations among themselves, resulting in a high standard deviation. This behavior can be attributed to the low shear resistance of the oxide coating and the influence of the ultrasonic process, which seemingly altered the microstructure of the base metal, suggesting an interference at the metal/oxide interface [13–15].

4.2 Fractographic Analysis of Aluminum and Composite Surfaces

In Figs. 2 and 3, the fractographies of the samples without PEO treatment and with PEO treatment are presented, respectively.

It can be observed that the fracture after the Lap Shear tests exhibited mixed failure modes, displaying cohesive failure (COH) in the thermoplastic matrix (yellow arrows)

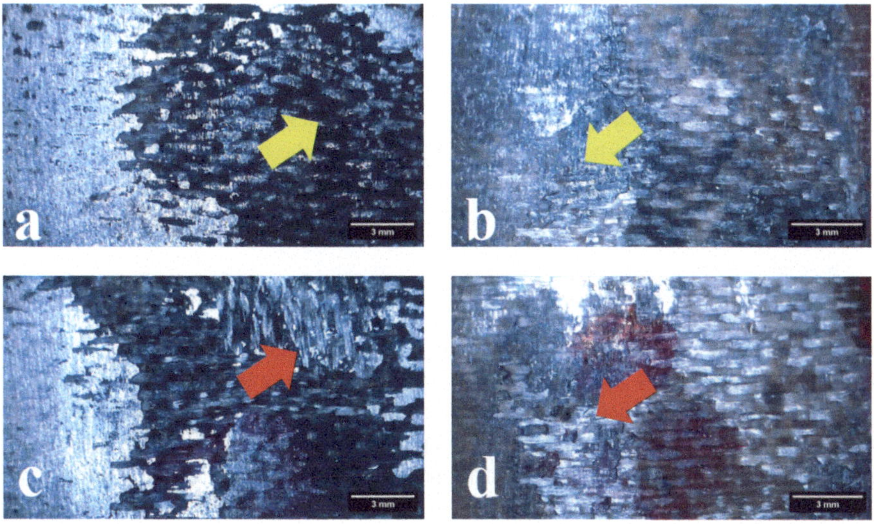

Fig. 2. Fractography of the samples without PEO treatment, with yellow arrows indicating cohesive failure and red arrows indicating light fiber rupture failure.

and light fiber rupture (FT), indicating that the adhesive strength between the aluminum and the thermoplastic matrix exceeds the internal strength of the polymeric material [16, 17]. The same applies to light fiber rupture, which can also be justified by the fact that, although the adhesion between the fiber and matrix is strong, the glass fibers may not fully transfer the loads, indicating potential areas for failure [18, 19].

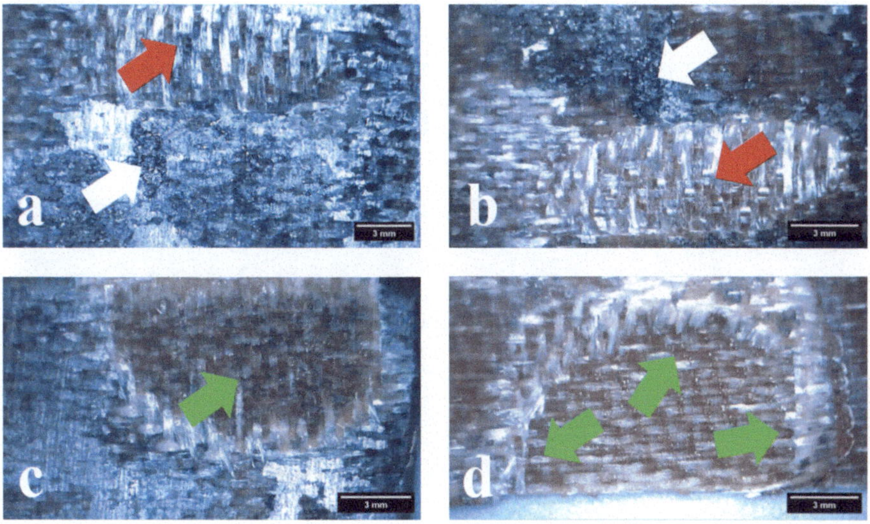

Fig. 3. Fractography of samples with PEO treatment, red arrows (light fiber rupture), white arrows (cohesive failure - PEO coating), and green arrows (fiber failure)

Upon analyzing the fracture of the samples treated by the PEO process, new modes of failure are observed, such as cohesive failure occurring within the oxide coating, indicated by the presence of oxide residues on the composite sample (Fig. 3b). This suggests that the oxide coating may have considerable thickness in this region, and combined with the ultrasonic process, it led to the failure of the coating. Additionally, fiber failure (FRP) occurs between the layers of the composite material (Fig. 3d), indicating greater interaction between the oxide coating, the matrix, and the fiber.

4.3 Scanning Electron Microscopy (SEM) of Plasma-Anodized AA2024

The comparison of scanning electron microscopy (SEM) images (Fig. 4) between untreated AA2024 alloy samples (A and B) and those treated with PEO (C and D) reveals significant differences in surface morphology that influence mechanical anchoring [22, 23]. The untreated samples exhibit a porous structure that enhances mechanical interlocking with the thermoplastic matrix, potentially leading to improved adhesion.

With the PEO treatment, the pores on the surface of the AA2024 alloy were sealed, which may hinder mechanical anchoring as there are no longer points for it; however, for corrosion analyses, this treatment can be beneficial as it acts as a barrier layer, preventing metallic interfaces (Al-Cu) from coming into contact with environments that may be rich in Cl^- ions [24, 25].

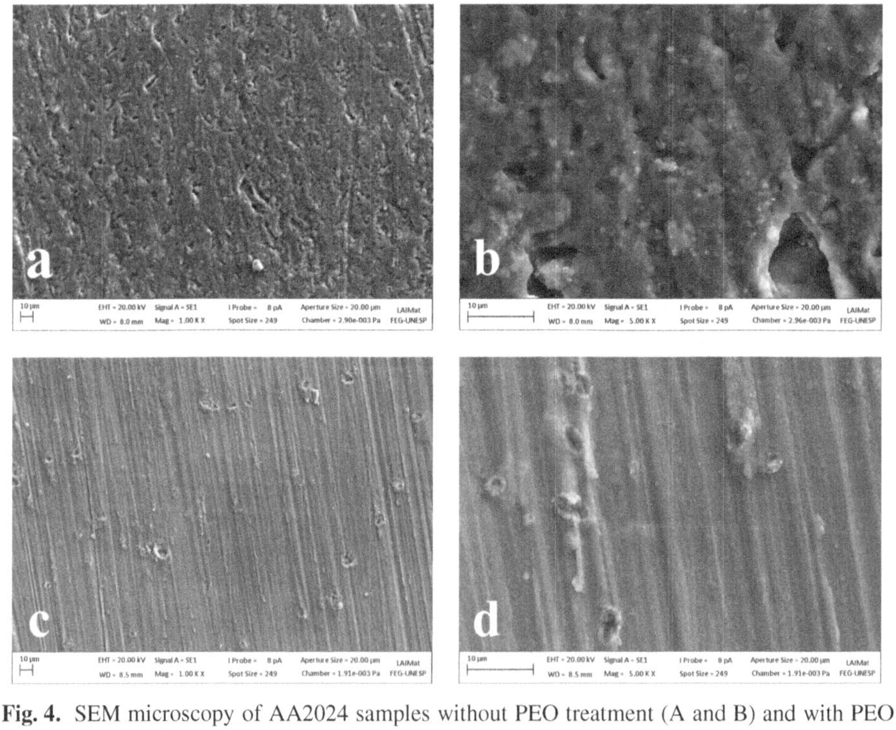

Fig. 4. SEM microscopy of AA2024 samples without PEO treatment (A and B) and with PEO treatment (C and D).

The presence of protrusions on the surface of the samples treated by the Plasma Electrolytic Oxidation (PEO) process is observed, with heights greater than and aligned scratches. These results suggest a supergrowth of the coating, and the observed scratches may have contributed to the delamination of the coating. Considering the ultrasonic welding process, it is inferred that there was a failure in this region with the thicker coating, leading to its breakdown, as noted by other authors [26, 27].

5 Conclusions

In this study, the AA2024 alloy anodized by Plasma Electrolytic Oxidation (PEO) was joined to the thermoplastic composite PEI-Glass Fiber through ultrasonic welding. The results indicated that the PEO treatment was not effective in promoting superior mechanical anchoring between the dissimilar materials, due to the low resistance of the oxide coating to shear stresses, as already reported in the literature. The samples treated by the PEO process showed a reduction in strength values of about 2.10%.

When analyzing the fracture regions between the samples, it was possible to identify mixed modes of failure. In the samples without PEO treatment, cohesive failures and minor fiber failures predominated, while in the samples treated with PEO, in addition to the minor fiber failures, cohesive failures in the oxide coating and fiber failures between the layers of the composite material were observed. These results indicate that the energy applied during the welding process may not have been distributed evenly, resulting in inconsistencies between the regions. Another hypothesis is that the oxide coating may have variable properties across the surface; if the area with fiber fracture were homogeneous, the treated samples would have demonstrated greater shear resistance.

Moreover, through electronic microscopy analysis, it was observed that the PEO process covered the existing pores of the untreated sample, a factor that was crucial in preventing mechanical interlocking between the polymer matrix of the composite and the aluminum alloy, although this may be beneficial for corrosion analysis of the AA2024 alloy.

Although the PEO coating enhances certain properties, its brittleness and the resulting mechanical challenges highlight the need for further optimization to improve performance under variable load conditions, which should be better assessed in future studies.

References

1. Lambiase, F., Scipioni, S.I., Lee, C.J., Ko, D.C., Liu, F.: A state-of-the-art review on advanced joining processes for metal-composite and metal-polymer hybrid structures. Materials **14**(8), 1890 (2021). https://doi.org/10.3390/ma14081890
2. Săftoiu, G.V., Constantin, C., Nicoară, A.I., Pelin, G., Ficai, D., Ficai, A.: Glass fibre-reinforced composite materials used in the aeronautical transport sector: a critical circular economy point of view. Sustainability **16**(11), 4632 (2024). https://doi.org/10.3390/su16114632
3. Gurenko, A.V.: Innovative applications of composite polymers in the automotive industry. Am. J. Interdisc. Innov. Res. **6**(05), 43–47 (2024). https://doi.org/10.37547/tajiir/Volume06Issue05-07

4. Ogundare, K.J., Ale, F., Yusuf, O.T., Olatoyinbo, S.F., Daniyan, I.A. Assessment of carbon fibre composites for product development in aerospace and transportation industries. SEB4SDG 1–8 (2024). https://doi.org/10.1109/SEB4SDG60871.2024.10629768
5. Jiang, B., Chen, Q., Yang, J.: Advances in joining technology of carbon fiber-reinforced thermoplastic composite materials and aluminum alloys. Int. J. Adv. Manuf. Technol. **110**, 2631–2649 (2020). https://doi.org/10.1007/s00170-020-06021-2
6. Francisco, D., Gonçalo, P.C., Arménio, N., Daniel, F.O., Pedro, M.G.P., Virgínia, I.: Joining of aluminum alloy AA7075 and titanium alloy Ti-6Al-4V through a friction stir welding-based process. Superalloys **13**(2), 249 (2023). https://doi.org/10.3390/met13020249
7. Tariq, A., Noor, Z.K., Arshad, N.S., Babar, A.: Investigations on surface properties and surface morphology in dissimilar joining of WE43 and AA-7075 alloy. Tribol. Int. (2024). https://doi.org/10.1016/j.triboint.2024.109302
8. Dumitrascu, D., Topala, I., Popa, G.: Dielectric barrier discharge technique in improving the wettability and adhesion properties of polymer surfaces. IEEE Trans. Plasma Sci. **33**(5), 1710–1714 (2005). https://doi.org/10.1109/TPS.2005.856335
9. Lucas, R.R., Mota, R.P., Abrahão, A.B.R.M., Botelho, E.C., Sales-Contini, R.C.M.: Characterization of oxide coating grown by plasma electrolytic oxidation (PEO) at different times on aluminum alloy AA2024-T3. MRS Commun. **12**, 266–271 (2022). https://doi.org/10.1557/s43579-022-00174-9
10. Lucas, R.R., et al.: Analysis of plasma electrolytic oxidation process parameters for optimizing adhesion in aluminum-composite hybrid structures. Appl. Sci. **14**(17), 7972 (2024). https://doi.org/10.3390/app14177972
11. Fernandez-Lopez, P., et al.: Data-driven optimization of plasma electrolytic oxidation (PEO) coatings with explainable artificial intelligence insights. Coatings **14**(8), 979 (2024). https://doi.org/10.3390/coatings14080979
12. ASTM (2019) D1002-10 - Standard test method for apparent shear strength of single-lap-joint adhesively bonded metal specimens by tension loading (Metal-to-Metal). https://www.astm.org/d1002-10r19.html
13. Gazenbiller, E., et al.: Computational damage modelling of PEO coated extruded magnesium tested in slow strain rate configuration. Surf. Coat. Technol. **446**, 128758 (2022). https://doi.org/10.1016/j.surfcoat.2022.128758
14. Nachtsheim, J., Ma, S., Burja, J., Markert, B.: In vitro evaluation of stress corrosion cracking susceptibility of PEO-coated rare-earth magnesium alloy WE43. Surf. Coat. Technol. **477**, 130391 (2024). https://doi.org/10.1016/j.surfcoat.2024.130391
15. Lucas, R.R., et al.: Characterization of the hybrid joint between AA2024-T3 alloy and thermoplastic composite obtained by oxy-fuel welding (OFW). AIMS Mater. Sci. **11**(3), 585–601 (2024). https://doi.org/10.3934/matersci.2024029
16. ASTM (2019) D5573-99 - Standard Practice for Classifying Failure Modes in Fiber-Reinforced-Plastic (FRP) Joints
17. Ejaz, H., Saboor, D., Sohail, M.S.B., Mansoor, A.H. Effect of various surface treatments on lap shear strength of aluminum adhesive joints. In: 19th International Bhurban Conference on Applied Sciences and Technology (IBCAST), pp. 76–81 (2022). https://doi.org/10.1109/IBCAST54850.2022.9990536
18. Qiao, Y., et al.: A study of adhesive bonding in metal–metal, metal–CFRP, and CFRP–CFRP material combinations under shear deformation: Fracture morphologies and damage mechanisms. Int. J. Adhes. Adhes. **127**, 103511 (2023). https://doi.org/10.1016/j.ijadhadh.2023.103511
19. Barros, R.B., Campilho, R.D., Sánchez-Arce, I.J., Dionísio, J.M., Madani, K.: Evaluation of the cracked lap shear test for mixed-mode fracture toughness estimation of adhesive joints. Proc. Inst. Mech. Eng. Part L: J. Mater. Des. Appl. (2022). https://doi.org/10.1177/14644207221140296

20. Zou, Y., et al.: In-situ SEM analysis of brittle plasma electrolytic oxidation coating bonded to plastic aluminum substrate: microstructure and fracture behaviors. Mater Charact **156**, 109851 (2019). https://doi.org/10.1016/j.matchar.2019.109851
21. Li, P., et al.: Effect of local 'over-growth' on fracture behaviors of coated titanium fiber fabricated by plasma electrolytic oxidation. Surf. Interfaces **36**, 102645 (2023). https://doi.org/10.1016/j.surfin.2023.102645
22. Akbarzadeh, S., Coelho, L.B., Dangreau, L., Lanzutti, A., Fedrizzi, L., Olivier, M.: Self-healing plasma electrolytic oxidation (PEO) coating developed by an assembly of corrosion inhibitive layer and sol-gel sealing on AA2024. Corros. Sci. **222**, 111424 (2023). https://doi.org/10.1016/j.corsci.2023.111424
23. Sopchenski, L., Robert, J., Touzin, M., Tricoteaux, A., Olivier, M.: Improvement of wear and corrosion protection of PEO on AA2024 via sol-gel sealing. Surf. Coat. Technol. **417**, 127195 (2021). https://doi.org/10.1016/j.surfcoat.2021.127195
24. Akbarzadeh, S., Sopchenski Santos, L., Vitry, V., Paint, Y., Olivier, M.: Improvement of the corrosion performance of AA2024 alloy by a duplex PEO/clay modified sol-gel nanocomposite coating. Surf. Coat. Technol. **434**, 128168 (2022). https://doi.org/10.1016/j.surfcoat.2022.128168
25. Rakoch, A.G., et al.: Role of cobalt additive on formation and anticorrosion properties of PEO coatings on AA2024 alloy in alkali-silicate electrolyte. Surf. Coat. Technol. **433**, 128075 (2022). https://doi.org/10.1016/j.surfcoat.2021.128075
26. Aliofkhazraei, M., et al.: Review of plasma electrolytic oxidation of titanium substrates: mechanism, properties, applications and limitations. Appl. Surf. Sci. Adv. **5**, 100121 (2021). https://doi.org/10.1016/j.apsadv.2021.100121
27. Attarzadeh, N., Ramana, C.V.: Plasma electrolytic oxidation ceramic coatings on zirconium (Zr) and ZrAlloys: Part I—growth mechanisms, microstructure, and chemical composition. Coatings **11**(6), 634 (2021). https://doi.org/10.3390/coatings11060634

Properties of Perlite-Based Geopolymer as a Thermal Insulation Material for Building Efficiency

Nina Polivoda[1], Mykola Melnychuk[1](✉), Barbara Kozub[2], Roman Ostapchuk[1], and Yaroslaw Babych[1]

[1] Lutsk National Technical University, 75, Lvivska Street, Lutsk 43018, Ukraine
m.melnychuk@lntu.edu.ua

[2] Cracow University of Technology, 37, Jana Pawła II Street, 31-864 Cracow, Poland

Abstract. Globally, the key issues related to technology and materials are the reasonable consumption of exhaustible resources, increasing technological efficiency, reducing waste, and sustainable production. The construction industry is responsible for huge carbon dioxide emissions and other harmful waste in this context. Moreover, it requires about 16 percent of the world's annual energy use for the production of cement-based products. Experts estimate that without effective changes, the negative impact on the environment and consumption will increase to 25% by 2050. This paper presents a geopolymer based on expanded perlite up to 85% that can be used as a modern lightweight thermal-insulation construction material. The results showed that the density of the samples varied from 330 kg/m3 to 510 kg/m3, the bending strength was from 1 MPa to 2 MPa, the compressive strength was from 13.5 MPa to 21 MPa, the thermal conductivity was from 0.09 W/m K to 0.130 W/m K, the water absorption coefficient was from 77% to 137%, and the total porosity of the material was from 60% to 65%. The results meet the requirements for lightweight building materials for wall construction regarding the leading indicators. Still, high water absorption significantly decreases strength and thermal conductivity after 100 h of the aging process. To reduce the water absorption of the perlite-based geopolymer, research is planned to apply a hydrophobic coating based on metakaolin.

Keywords: Expanded Perlite · Mechanical Properties · Thermal Conductivity · Lightweight Masonry Block · Perlite-Based Geopolymers

1 Introduction

The state of energy production and consumption in Ukraine has always determined its political situation and economic growth. Today, energy independence is a matter of national security and sustainable economic development. Due to limited energy resources, Ukraine imports about 50% of its total energy needs.

The Ukrainian government has taken certain steps in the last decade to reduce energy dependence and is striving to increase the share of renewable and alternative energy.

F. J. G. Silva et al. (Eds.): DSMIE 2025, LNME, pp. 76–87, 2025.
https://doi.org/10.1007/978-3-031-96413-8_8

Another effective method is to reduce the energy consumption ratio. In recent years, there has been a growing interest in energy conservation, as energy production is mainly achieved by burning fossil fuels, which will run out shortly. Today, the world's total primary energy consumption has reached approximately 85% from fossil fuels such as coal, oil, and natural gas [1, 2]. Approximately 50% of this consumption is spent directly on buildings. The EU is trying to abandon gas and oil as energy sources that pose both environmental and economic, military, and political risks. Thus, reducing energy consumption, especially in buildings, is crucial, as it will positively impact economic growth in the short and long term. Ukraine consumed 2.4 times more energy than the global average and 3.1 times more than the European Union. Since modern civilization cannot completely abandon energy consumption, the only way out of this situation is for it to rationally consume based on using renewable energy sources, energy-saving energy sources, energy saving, and the introduction of energy-efficient technologies.

As for materials, approximately 10 billion tons of structural materials are produced annually globally. This volume of production is responsible for 16% of global energy consumption and carbon emissions, with cement production alone accounting for almost 5%. So, summarizing the above, developing alternative thermal insulation construction materials is a relevant and essential task [3–5].

2 Literature Review

Many studies have been published aimed at producing heat-insulating materials from expanded perlite with low thermal conductivity. In their scientific work, Uller et al. experimentally investigated the production of thermal insulation panels based on expanded perlite with the lowest thermal conductivity at different cement-based binder contents. The authors found that the thermal conductivity values of thermal insulation panels decrease depending on the decrease in the binder ratio in the mixture. However, the reduction of the binder had a negative impact on the strength of the panels [6, 8]. In work [7], cementless pastes and mortar were produced based on alkaline activation using sodium hydroxide and sodium silicate of raw perlite with a hydrogen peroxide blowing agent. Articles [8, 9] represented the results of studies of geopolymers with the addition of expanded perlite, but the compressive strength was insignificant. The article presents the results of a study of the effect on thermal conductivity and compressive strength of geopolymers based on metakaolin with fillers in the form of swollen perlite and cellulose fibers. The strength was found not to exceed 6 MPa, and the thermal conductivity is in the region of 0.10 W/m [10]. In early works, perlite was considered a source of aluminum silicates in the form of a filler or precursor. Still, according to research results, it was considered an inefficient raw material with low mechanical properties in the final products [11–15]. For instance, in the study by Vance et al. [14], the results of the reactivity of perlite of different fractions with NaOH solution were presented. The geopolymerization temperature of the composite was determined. However, the composition based on perlite alone did not provide sufficient strength, and only the addition of fly ash to the composition increased the compressive strength. It was investigated that perlite waste in interaction with alkaline solutions (10 M NaOH) formed crystalline zeolite 4A and other crystalline phases. However, the compressive strength of the obtained geopolymers was low. After the addition of perlite up to 22.5%, fly ash up to 40%, and calcined kaolin clay 10% by weight, it was possible to increase the mechanical properties to acceptable values [16–19] It was revealed [20, 21] that the formation of a zeolite phase and a heterogeneous structure in perlite geopolymers causes a decrease in strength. The maximum ratio of perlite to geopolymer at which the mechanical properties are reasonable for the production of materials is in the range of 70–80%.

Papadopoulos [22] emphasizes that despite a significant amount of research on materials based on expanded perlite in the last decade, the thermal properties of such composites have not improved significantly, except for fire resistance. The study of pressure-sintered composites based on expanded perlite is considered by the authors in [23]. In the study by Kornienko et al. (2023), expanded perlite and vermiculite were used as lightweight fillers in geopolymers, the authors found a slight effect of fillers on strength properties, but the density and thermal conductivity of materials decreased significantly. Such fillers are characterized by good adhesion to the geopolymer matrix [24]. The filling of the geopolymer based on fly ash with 10% expanded perlite led to an increase in plasticity but a decrease in compressive strength by up to 40% [25].

Based on the above, the main goal of this work was to develop a geopolymer for structural purposes with a high degree of filling up to 85% by weight of expanded perlite as a filler, which provides a low coefficient of thermal conductivity. According to the analysis of literary sources, the authors failed to obtain sufficient compressive strength with a perlite filling content above 50% by weight without the use of metakaolin as a binder. Accordingly, the developed materials as an alternative to materials based on cement technologies should meet the general requirements for structural and thermal insulation building materials.

3 Research Methodology

The following materials were used to prepare the compositions: expanded perlite from the Fogosh deposit, sieved on laboratory sieves 1.8 to 2.2 mm, liquid sodium silicate glass as a binder, and plasticizer Sika Mix Plus (Switzerland).

The components were added in the following sequence: first, the plasticizer was mixed with perlite, then the binder was added. To form the compositions, metal molds with dimensions of 50 × 50 × 50 mm were used, which were pre-lubricated with graphite grease, then the compositions in the mold were pressed with a force of 980 N and sintered in an oven at 200 °C for 1 h.

Five compositions of the composite were prepared, the general view of which is shown in Fig. 1: No. 1 - (5 mass fraction of perlite +1 mass fraction of liquid glass); No. 2 - (5 mass fraction of perlite +2 mass fraction of liquid glass); No. 3 - (5 mass fraction of perlite +3 mass fraction of liquid glass); No. 4 - (5 mass fraction of perlite +4 mass fraction of liquid glass); No. 5 - (5 mass fraction of perlite +5 mass fraction of liquid glass) and 0.1 mass fraction plasticizer was added to each composition. The ratio of the components was determined based on the results of mathematical modeling using MATLAB tools.

The water absorption of the samples (Wi) was determined as the average of the water absorption of all samples, calculated to an accuracy of 1%. The samples were placed in

Fig. 1. General view of samples.

a bath on a mesh stand and loaded with a weight. Thermal conductivity measurements were performed using a plating apparatus HZM 446 Lambda Small (NEZTSCH, Selb, Germany). The temperature stabilization accuracy was ±0.2 °C.

Porosity measurements were performed by using a Quantachrome Instruments Poremaster mercury porosimeter. The study was conducted for composites No. 1, No. 3, and No. 5, each subjected to high- and low-pressure tests. The maximum value during the low-pressure test is about 50 PSI; during the high-pressure test, this value is about 35.000 PSI, due to the difference in the volume of mercury introduced into the sample. During a high-pressure test, mercury may have penetrated less accessible pores; therefore, the amount of mercury injected is greater. Mercury porosimeter can measure pores with a size of approximately 250–0.003 μm.

The compressive strength test was carried out on the MATEST 3000 km device. The linear rate of force increase was 1.25 kN per second. All tests were carried out after 28 days of sample curing. All data are the average of three to six repetitions.

The elemental and oxide composition was measured using two methods (for comparative purposes)—the first method used loosely packed expanded perlite powder for measurements. In the second method, measurements were performed on compressed material by using The S2 PUMA Series 2 XY Autochanger X-ray fluorescence spectrometer (XRF) (Bruker, Billerica, Volumeachusetts, United States To prepare the sample for measurement, 6 g of expanded perlite was weighed, to which 1.5 g of wax (Licowax® C micro powder, PD Instruments, Toszek, Poland) was added.

The microstructure observations of expanded perlite were performed using a JEOL JSM-820 scanning electron microscope (IXR Inc., Austin, TX, USA). Before observation, the samples were prepared by sputtering with a gold layer in a JOEL DII-29030SCTR vacuum coater (IXR Inc., Austin, TX, USA).

The perlite particle size distribution was analyzed using a laser particle analyzer, Anton Paar GmbH (Graz, Austria), which uses the light diffraction technique. The measurement was performed in wet mode, which required the preparation of a metakaolin sample and its introduction into the device, where the light scattering analysis was then performed. The study established statistical values of grain size based on six measurements.

4 Results and Discussion

4.1 Characterization of Raw Material

The chemical composition of the tested precursors was determined by the XRF method. Perlite is considered an important raw material for the production of geopolymers; however, the reactivity depends on the chemical composition. The results of the chemical composition study are given in Table 1.

Table 1. Perlite chemical element compositions (X-ray fluorescence).

Element [Volume%]											
	Si	Al	K	Fe	Na	Ca	Ti	Mg	Cl	Mn	Rb
Expanded perlite	65.73	11.95	10.65	3.96	3.56	3.08	0.23	0.22	0.18	0.15	0.07

The perlite particles had an irregular spherical morphology with a rough and porous texture, which is an advantage for achieving a successful geopolymerization process (Fig. 2).

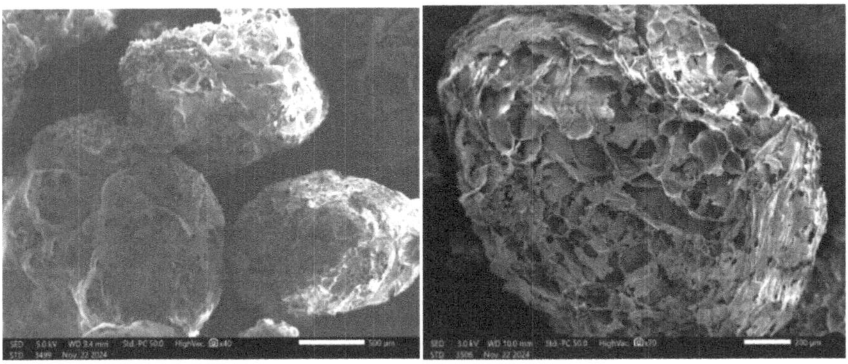

a b

Fig. 2. SEM micrographs of expanded perlite morphology: a) ×40; b) ×70.

This improves the rheological properties of the mixture, increasing its ease of placement. In addition, it reduces the need for liquids and has a beneficial effect on the mechanical properties of geopolymers [26]. The size of the perlite particles (Fig. 3) after sieving ranged from 10 to 2000 μm, where 90% of the particles are larger than 700 μm and the D_{50} distribution width is 650 μm. The larger particle size results in a lower

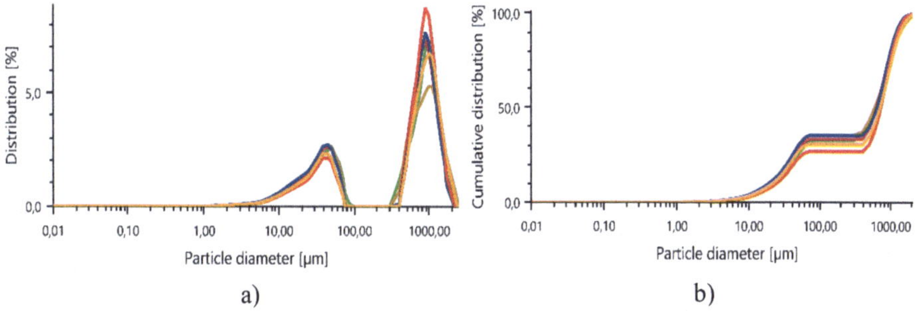

Fig. 3. Results of particle size analysis: (a) particle size distribution histogram; (b) cumulative particle size distribution curves.

specific density and greater porosity, which positively affects the thermal conductivity coefficient.

Furthermore, the porosity is the biggest at the biggest particle size and the voids can be better filled within big particles, which can lead to denser and stronger geopolymer products. Figure 4 shows the diffractogram recorder for expanded perlite. The results of X-ray studies showed that in the analyzed perlite, it was possible to detect the presence of impurities in the form of the SiO_2 crystalline phase. This phase does not participate directly in the polymerization process. The recorded XRD curve indicates the presence of the amorphous phase in the tested perlite (a very large hump). Amorphous phases play an important role in the polymerization process [25–28, 29] (Table 2).

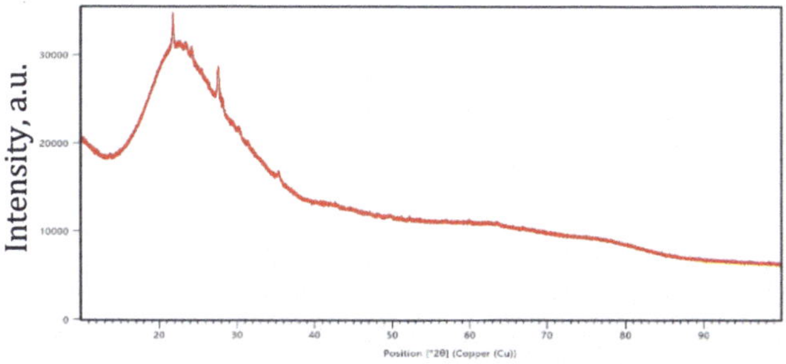

Fig. 4. Diffractogram for perlite

Furthermore, the porosity is the biggest at the biggest particle size and the voids can be better filled within big particles, which can lead to denser and stronger geopolymer products. Figure 4 shows the diffractogram recorder for expanded perlite. The results of X-ray studies showed that in the analyzed perlite, it was possible to detect the presence of impurities in the form of the SiO_2 crystalline phase. This phase does not participate directly in the polymerization process. The recorded XRD curve indicates the presence of the amorphous phase in the tested perlite (a very large hump). Amorphous phases play an important role in the polymerization process [24–28, 29].

Table 2. Particle size distribution in expanded perlite.

Name	D_{10}(μm)	D_{50}(μm)	D_{90}(μm)	Mean size(μm)	Span	Color
Perlite 1	22.183	646.033	1272.342	651.459	1.935	
Perlite 2	19.650	589.593	1372.794	652.600	2.295	
Perlite 3	19.967	629.843	1210.398	627.373	1.890	
Perlite 4	18.586	635.847	1192.769	615.257	1.847	
Perlite 5	23.525	730.627	1265.677	712.669	1.700	
Perlite 6	21.011	708.141	1363.656	712.337	1.896	

4.2 Water Absorption Test

The test for water absorption of a perlite-based thermal insulation composite consists in measuring the volume of water absorbed by a dry material sample completely immersed in water for a given time, presented in Table 3.

Table 3. Water absorption and density thermal insulation composite.

No. of composition	The ratio (perlite + liquid glass)	Water absorption, Wi, %	Density, kg/m^3
1	5:1	136.2	330
2	5:2	120.4	360
3	5:3	98.8	410
4	5:4	84.7	459
5	5:5	76.5	510

It was established that the minimum value of water absorption (Wi = 76.5%) with a density (510 kg/m^3) is obtained for composition No. 5, which has a ratio of 5:5 (perlite + liquid glass) and was formed at a temperature of 400 °C. The highest value of water absorption (Wi = 136.2%) with a density (of 330 kg/m^3) is observed in composition No. 1, which has a ratio of 5:1 (perlite + liquid glass) According to the dependence, water absorption decreases with an increase in the amount of liquid glass, which can be explained as a consequence of a decrease in the number of open and closed pores, as well as a reduction in the area of perlite inclusions due to the coating of the liquid glass matrix, which in turn blocks the contact of water with the adsorbing surface of perlite, which is a water-absorbing component in the composition. According to the results, the material requires a surface coating with a hydrophobic material to meet the operational requirements in real conditions.

Table 4. The results of the porosity measurements.

Sample	Total interparticle porosity [%]	Total intraparticle porosity [%]	Total porosity [%]	Theoretical porosity [%]
1 Low	65.14	0.000	65.14	79.31
1 High	0.011	0.0156	0.0266	77.68
3 Low	64.4	0.000	64.4	74.63
3 High	0.005	0.025	0.030	76.25
5 Low	61.55	0.000/-	61.55	70.6
5 High	0.027	0.084	0.111	72.7

4.3 Porosity Measurement

The summation of the two indicators provides information on the total porosity of the materials under test. Moreover, the theoretical porosity of the material is calculated based on the measurement. The results of the porosity measurements are presented in Table 4.

Testing under high pressure provides information on intraparticle porosity. The value depends on material characteristics, especially pore characteristics. In the case of closed pores, information on porosity may be missing. The ability to conduct tests under high- and low-pressure conditions provides more objective information.

Based on the results obtained, it was found that the theoretical porosity of perlite-based geopolymer, depending on the content of the perlite, is about 79% for composition No. 1, about 75% for composition No. 3, and about 71% for composition No. 5.

4.4 Mechanical Properties

Compressive strength test results are shown in Fig. 5. The highest strength value is 23 MPa for the sample of thermoperlite No. 5 with a component ratio of (5:5). According to the obtained dependence, the strength of the composite increases almost linearly with an increase in the binder content up to composition No. 3, and then no significant increase in strength is observed.

The obtained strength values make it possible to use this material in construction since the widely used ceramic brick of the M150 brand has a compressive strength of 15 MPa. Compared to other alternative materials, such as those used by the authors of [24], the strength of the geopolymer filled with perlite was only 2.3 MPa, and that filled with vermiculite was 3.75 MPa. The obtained strength values are higher than the average values obtained by other authors [6, 11, 23]. Mechanical properties correlate with the porosity of the material and its density.

Thus, it is highly probable that it results from water saturation and disruption of the formed relationships between the binder and filler, by comparing the obtained compressive strength results with those of [13, 14]. We can state that for the created geopolymers, even at a specific density of 300 kg/m^3, the compressive strength meets the standard's requirements for such materials. A low thermal conductivity coefficient was obtained, which is one of the most essential properties. The strength results from forming good

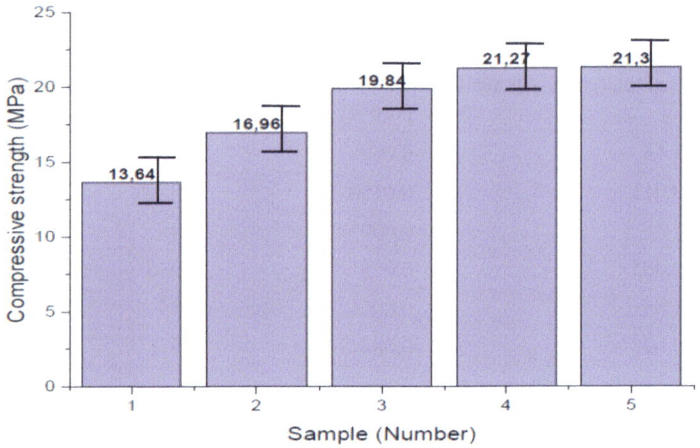

Fig. 5. The results of compressive strength tests (after 28 days of curing).

chemical bonds between the particles due to temperature treatment under load. Similar materials based on metakaolin studied in [28] have higher strength values but are inferior in terms of thermal conductivity, and metakaolin is, on average, 7 times more expensive than perlite.

4.5 Durability and Thermal Conductivity Test

The samples were moistened to the humidity values of the operating conditions hermetically sealed in special plastic bags and exposed to different temperatures. Thus, at the first stage of the test, the samples were subjected to freezing at a temperature of −22 °C. Then, the samples were thawed at a temperature of +20 °C, and then, in the third stage, they were heated to a temperature of +60 °C. After every tenth cycle, samples were taken to determine their thermal conductivity under standard conditions, the results of which are shown in Fig. 6, and changes in the appearance of the material were recorded.

According to the results of the visual inspection of the test samples after 100 cycles of climatic effects of freezing-thawing-heating, it was found that the appearance of the samples of the thermal insulation composition does not change - the change in the geometric dimensions of the samples is within the limits of permissible values, there is no change in the color and structure of the material, the general view of the composition is shown in Fig. 7.

The tests were performed following DSTU B V.2.7-182-2009. As a result, it was determined that the period of usage of perlite geopolymer for thermal insulation of buildings is at least 50 years. This indicator not only fully complies with the norms of DSTU B.2.6-31: 2006, but also exceeds it by 2 times.

Analyzing the study results, the specific density is in the range of 330–510 kg/m^3 with a total porosity of 60 to 65%, compressive strength from 13 to 20 MPa, and thermal conductivity from 0.09 to 0.13. W/(m K). It can be stated that the experimental materials meet the requirements of Ukrainian standards in terms of the leading indicators of mechanical characteristics, as they are not inferior to widespread Autoclaved Aerated

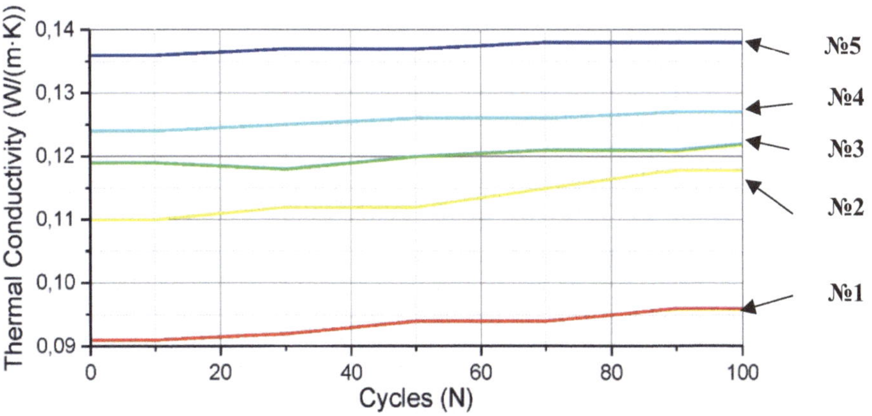

Fig. 6. Relation of thermal conductivity to the number of thermal cycles.

Concrete (AAC) and have significantly better thermal conductivity and specific density compared to bricks.

However, according to the results, significant drawbacks need to be addressed in future studies, such as the high-water absorption coefficient, which leads to a decrease in mechanical properties, as found during the wetting aging test. In addition, the sintering molding method requires additional energy consumption. Future research will focus on molding methods without additional sintering and reducing water absorption.

5 Conclusions

This paper presents the possibility of the development of lightweight masonry blocks based on expanded perlite. It has been shown that can produce geopolymer with content up to 85% expanded perlite, which was formed by treatment with pressure and temperature of 200 °C.

According to the results of the durability test, it was found that the phase of use of perlite-based geopolymers for thermal insulation of buildings is at least 50 years. After 100 thermal cycles of freezing and heating, the thermal conductivity of the material increased by an average of 10% for almost all compositions. It was found that the total porosity, depending on the composition, ranges from 65% to 60%, which correlates well with the results of determining the thermal conductivity, which is in the range of 0.09 to 0.13 W/(m K). The water absorption coefficient was from 77% to 137%, The compressive strength of the studied composites depends on the composition and ranges from 13 to 21 MPa, the bending strength was from 1 MPa to 2 MPa. It has been investigated that this material does not contain harmful components and is eco-friendly.

Accordingly, this material, based on the results obtained, has prospects as a lightweight construction building material. In further studies, it is necessary to find the methods for reducing water adsorption and investigate the impact strength and fire resistance.

References

1. British Petrol (2018). https://www.bp.com/en/global/corporate/energy-economics/statistical-review-of-world-energy/primary-energy. Accessed 02 Nov 2018
2. U.S. Energy Information Administration (2018). https://www.eia.gov/beta/international/analysis.php?iso=TUR. Accessed 02 Nov 2018
3. Arrigoni, A., et al.: Life cycle greenhouse gas emissions of concrete containing supplementary cementitious materials: cut-off vs. substitution. J. Clean. Prod. **263**, 121465 (2020). https://doi.org/10.1016/j.jclepro.2020.121465
4. Gatt, D., Yousif, C., Cellura, M., Camilleri, L., Guarino, F.: Assessment of building energy modeling studies to meet the requirements of the new Energy Performance of Buildings Directive. Renew. Sustain. Energy Rev. **127**, 109886 (2020). https://doi.org/10.1016/j.rser.2020.109886
5. Łach, M., Kozub, B., Bednarz, S., Bąk, A., Melnychuk, M., Masłoń, A.: The influence of the addition of basalt powder on the properties of foamed geopolymers. Materials **17**, 2336 (2024). https://doi.org/10.3390/ma17102336
6. Uluer, O., et al.: An investigation of usability of expanded perlite in heat insulation technologies. Pamukkale Univ. J. Eng. Sci. **24**(1), 36–42 (2018). https://doi.org/10.5505/pajes.2017.61687
7. Mehmet, C.A., Ali, İ.Ç., Ramazan, K., Ahmet, Ş., Nebi, Ö., Yasin, O.Ö.: Production of perlite-based-aerated geopolymer using hydrogen peroxide as an eco-friendly material for energy-efficient buildings. J. Mater. Res. Technol. **24**, 81–99 (2023), ISSN 2238–7854, https://doi.org/10.1016/j.jmrt.2023.02.179
8. Huan, G., et al.: Wang: Improvement of durability of porous perlite geopolymer-based thermal insulation material under hot and humid environment. Constr. Build. Mater. **313**, 125417 (2021). https://doi.org/10.1016/j.conbuildmat.2021.125417
9. Rashad, A.M., Khalil, M.H., El-Nashar, M.H.: Insulation efficiency of alkali-activated lightweight mortars containing different ratios of binder/expanded perlite fine aggregate. Innov. Infrastruct. Solut. **6**, 156 (2021). https://doi.org/10.1007/s41062-021-00524-x
10. Kurek, I., et al.: Foamed eco-geopolymer modified by perlite and cellulose as a construction material for energy-efficient buildings. Energies **15**, 4297 (2022). https://doi.org/10.3390/en15124297
11. Mohabbi, M.: Investigation of sulfates effects in perlite-based geopolymer. Struct. Concr. **20**, 1402–1410 (2019). https://doi.org/10.1002/suco.201900039
12. Çelikten, S., Işıkdağ, B.: Properties of geopolymer mortars derived from ground calcined perlite and NaOH solution. Eur. J. Environ. Civ. Eng. **27**(9), 2907–2921 (2021). https://doi.org/10.1080/19648189.2021.1879939
13. Erdogan, S.T.: Properties of ground perlite geopolymer mortars. J. Mater. Civil Eng. **27** (2015), https://doi.org/10.1061/(ASCE)MT.1943-5533.0001172
14. Vance, E.R., Perera, D. S., Imperia, P., Cassidy, D.J., Davis, J., Gourley, J.T.: Perlite waste as a precursor for geopolymer formation. J. Aust. Ceramic Soc. **45**, 44–49 (2009). http://apo.ansto.gov.au/dspace/handle/10238/3130
15. Laukhin, D., Ziborov, K., Fedoryachenko, S., Rott, N.: Influence of temperature-strain parameters of shelters reinforcing materials on increasing properties in the Z-direction. In: Materials Science Forum, vol. 1126, pp. 111–118 (2024). https://doi.org/10.4028/p-j10tbj
16. Baran, P., Nazarko, M., Włosińska, E., Kanciruk, A., Zarębska, K.: Synthesis of geopolymers derived from fly ash with an addition of perlite. J. Clean. Prod. **293**, 126112 (2021). https://doi.org/10.1016/j.jclepro.2021.126112
17. Aziz, A., et al.: Effect of acidic volcanic perlite rock on physio-mechanical properties and microstructure of natural pozzolan based geopolymers. Case Stud. Constr. Mater. **15**, e00712 (2021). https://doi.org/10.1016/j.cscm.2021.e00712

18. Voottipruex, P., Teerawattanasuk, C., Sramoon, W., Meepon, I.: Stabilization of soft clay using perlite geopolymer activated by sodium hydroxide. Int. J. Geosynth. Ground Eng. **8**(1), 1–8 (2022). https://doi.org/10.1007/s40891-022-00350-w
19. Szabó, R., Dolgos, F., Debreczeni, A.: Characterization of mechanically activated fly ash-based lightweight geopolymer composite prepared with ultrahigh expanded perlite content. Ceram. Int. **48**, 4261–4269 (2022). https://doi.org/10.1016/j.ceramint.2021.10.218
20. Saufi, H., El Alouani, M., Alehyen, S., El Achouri, M., Aride, J., Taibi, M.H.: Photocatalytic degradation of methylene blue from aqueous medium onto perlite-based geopolymer. Int. J. Chem. Eng. **2020**(1), 9498349 (2020). https://doi.org/10.1155/2020/9498349
21. Melnychuk, M., Poteichuk, M., Kashytskyi, V., Sosnowski, M., Kutsyk, S.: Thermo-mechanical properties of perlite composite. In: Tonkonogyi, V. (ed.) Advanced Manufacturing Processes II: Selected Papers from the 2nd Grabchenko's International Conference on Advanced Manufacturing Processes (InterPartner-2020), September 8-11, 2020, Odessa, Ukraine, pp. 330–338. Springer International Publishing, Cham (2021). https://doi.org/10.1007/978-3-030-68014-5_33
22. Papadopoulos, A.M.: State of the art in thermal insulation materials and aims for future developments. Energy Build. **37**(1), 77–86 (2005). https://doi.org/10.1016/j.enbuild.2004.05.006
23. Korniejenko, K., Pławecka, K., Bulut, A., Şahin, B., Azizağaoğlu, G., Figiela, B.: Deveopment of lightweight geopolymer composites containing perlite and vermiculite. Jo. Achiev. Mater. Manuf. Eng. **117**(2), 49–56 (2023). https://doi.org/10.5604/01.3001.0053.6696
24. Baran, P., Nazarko, M., Włosińska, E., Kanciruk, A., Zarębska, K.: Synthesis of geopolymers derived from fly ash with an addition of perlite. J. Clean. Prod. **293**, 126112 (2021). https://doi.org/10.1016/j.jclepro.2021.126112
25. Korniejenko, K., Łach, M., Marczyk, J., Ziejewska, C., Halyag, N.P., Mucsi, G.: Fly ash as a raw material for geopolymerisation mineralogical composition and morphology. IOP Conf. Ser. Mater. Sci. Eng. **706**, 1–8 (2019). https://doi.org/10.1088/1757-899X/706/1/012006
26. Kumar, S., Mucsi, G., Kristály, F., Pekker, P.: Mechanical activation of fly ash and its influence on micro and nano-structural behavior of resulting geopolymers. Adv. Powder Technol. **28**(3), 805–813 (2017). ISSN 0921–8831. https://doi.org/10.1016/j.apt.2016.11.027
27. Korniejenko, K., Halyag, N.P., Mucsi, G.: Fly ash as a raw material for geopolymerization-chemical composition and physical properties. IOP Conf. Ser. Material. Sci. Eng. **706**, 012002 (2019). https://doi.org/10.1088/1757-899X/706/1/012006
28. Pavlík, V., Bisaha, J.: Lightweight mortars based on expanded perlite. In Key Engineering Materials, vol. 776, pp. 104–117. Trans Tech Publications, Ltd. (2018). https://doi.org/10.4028/www.scientific.net/kem.776.104

Production and Surface Research of a New Material with the Addition of Diatomite for Potential Filtration Applications

Oleksandr Povstyanoy(✉), Nataliia Imbirovych, Bohdan Smetjukh, Yuliia Karbovska, and Dmytro Seleznov

Lutsk National Technical University, 75, Lvivska Street, Lutsk 43018, Ukraine
povstjanoj@ukr.net

Abstract. This work is dedicated to the research and study of the possibility of using Natural minerals such as Diatomaceous Earth (diatoms) for filtering various environments. The extraordinary progress and prospects of diatomaceous earth (DE) as a biomatrix material for creating porous materials are relevant and interesting. The unique structures of DE, including its mesoporosity and large surface area, have made it a key material for the filtration process. Its biotemplating ability can initiate the creation of highly efficient porous permeable materials. Its sufficient natural abundance and ecological properties are consistent with the increasing emphasis on sustainability in environmental solutions. Its complex structures and biotemplating capabilities will pave a compelling path in the future for the advancement of sustainable, highly efficient solutions for filtering environments, marking a significant step towards a more ecological and efficient future for humanity.

This research paper explores the potential of DE as a bio-matrix material for creating a permeable material without delving into the complex filtration processes. Previous studies have shown that the use of Diatomaceous Earth with titanium powder in various proportions allows us to talk about the possibility of creating highly efficient porous permeable materials with branched porosity even for drinking water filtration.

Keywords: Diatomaceous Earth · porous materials · pressing

1 Introduction

Filtration - the process of separating suspensions or aerosols using filter partitions that pass liquid or gas, but retain solid particles, is carried out in special devices - filters.

The effectiveness of filter production in the world in modern economic conditions depends significantly on the rational selection of cheap materials and the use of resource-saving technologies. The competitiveness of industrial corporations, enterprises, factories, and firms is formed by regulating the cost of production and using the latest technologies. Modern technological innovations allow considering natural materials as an additional source of raw materials.

F. J. G. Silva et al. (Eds.): DSMIE 2025, LNME, pp. 88–98, 2025.
https://doi.org/10.1007/978-3-031-96413-8_9

In materials science, new ideas often arise from inspiration and study of natural design works. Structured surfaces are an integral part of many fields of science, including materials science, namely chemistry, microelectronics, optics, photonics, filtration, etc. A key innovation here is the use of bio-templated materials, in particular those made from Diatomaceous Earth.

Natural minerals such as Diatomaceous Earth (diatoms) are widely used in many areas of the national economy, are used in various industries, namely in filtering for various types of water. The use of this material will help to solve a number of problems related to the filtration and purification of wastewater and drinking water, as Diatomaceous Earth has high operational properties due to the uniqueness of its structure.

In modern water purification technology, when drinking water is consumed, the use of natural and artificial materials is expected to solve the problems. Basically, sorption processes are implemented using activated carbon or their analogues - graphite mineral sorbents, for the same purposes saponites, zeolites, clinoptilolites, etc. are used as filter material or contact loading.

Diatomaceous earth filtration is a process that uses diatoms, or diatomaceous earth, the skeletal remains of small single-celled organisms, as a filter medium. This type of diatoms filtration relies on a layer of diatomaceous earth placed on top of a filter element or baffle [1, 2].

This filtering process is expensive, difficult to use and not always economically beneficial.

Powder metallurgy with each newly developed technological process demonstrates advantages that allow obtaining materials with better or completely new properties, or manufacturing products in the most economically beneficial way.

Therefore, the creation of new filters using Diatomaceous Earth by powder metallurgy methods is relevant at the current stage of industrial development.

2 Literature Review

Due to the growing need to find alternative raw materials for filtration, the demandfor natural resources has increased in recent years [3].

To solve these problems, in recent years there has been a growing interest in finding solutions that aim to urgently reduce the negative impact associated with environmental pollution, which, in turn, is one of the biggest challenges facing modern society.

The importance of porous ceramics cannot be underestimated, as its unique properties such as high melting point, high corrosion resistance, high wear resistance, low density, low thermal conductivity and low dielectric constant are extremely important for the filtration process.

In particular, porous ceramic materials are possible materials for creating porous filter ceramics. The driving force behind the development of porous ceramic filters with a higher level of thermal and chemical stability is the availability of high-quality starting ceramic material [4].

In general, the most important aspects of porous ceramic filters are their permeability. Therefore, it is important to control the average and largest pore size while maintaining the degree of permeability [5].

The characteristics of the pores - depth, tortuosity, closedness and openness - are difficult in the application of ceramic filters [6]. Therefore, the main attention is paid to various approaches that allow controlling the microstructural features of the porous layer of the filter material, which ultimately determine the permeability properties.

Despite the fact that the methods of obtaining porous filter ceramics are widely described in literature sources [7, 8], the relationship between the characteristics of pores and the filter properties of porous permeable materials has not yet been definitively established.

In addition, ceramic filter materials are usually composites consisting of several layers of one or more different ceramic materials or a mixture thereof. Therefore, porous ceramic filters are usually manufactured in several stages.

The application and separation mechanisms of porous ceramic materials correspond to the pore size of the filters. Since it is difficult to create a surface layer of microporous or dense materials due to their low penetration characteristics, it is worth investigating porous ceramic materials consisting of only one layer and a surface layer consisting of the same macroporous material.

Recent developments in porous ceramic materials have increased the need to investigate the surface layer, as the overall permeability and reliability of the pores in the surface layer are critical for the application of porous ceramic filters.

Although various literature sources [9–11] investigate commonly used materials for ceramic filters, such as Al_2O_3, TiO_2, ZrO_2, SiO_2 and composites of these materials, few scientific studies have been conducted on the application of starting particles for - lack of ceramic porous materials, such as diatomite [12].

Diatomite is a sedimentary rock derived from the siliceous fossilized skeletons of diatoms, which are composed of hard cell walls called frustules. To date, no detailed studies have been published on the use of a surface layer of ceramic porous material made of diatomite or its composite.

Diatomaceous earth, or diatomite, fits into this context, since it is a material consisting of 87–91% SiO_2 [13], a large part of which (25–100%) is usually amophous, namely opal silica (opal-A). Due to the high content of amorphous silica and the entire surface of diatomite, it has a high filtration capacity for use in the production of porous cement-based composites [14, 15].

Therefore, the purpose of this research was to create and study the structure of a new material based on diatomite with the prospect of developing new porous permeable materials for drinking water filtration on this basis.

3 Research Methodology

The analysis of many literary sources shows that promising methods of obtaining filters are methods aimed at creating filters with an anisotropic structure, in which the size and number of pores change in the direction of filtration. In such filters, the fineness of the filtration will be determined by the layer with the minimum pore size, the permeability is an integral value determined by the porous structure of the entire material, and the pollutant during the filtering process is distributed over the entire volume of the filter element, which allows to increase the service life of the filter.

The selection of charge materials is of great importance for obtaining high-quality products [16].

In this work, the basis of the composite components of the charge for obtaining filter partitions was Diatomaceous Earth powder and titanium. Since titanium is characterized by an interesting combination of such properties as high strength, low density, resistance to corrosion and biocompatibility. Therefore, titanium is increasingly used in industries with high requirements, such as the filtration industry, the production of biomedical devices, etc.

The batch composition of the components was determined experimentally.

The ratio of the charge components is summarized in Table 1.

Table 1. Ratio of charge components.

№ sample	Diatoms, %	Ti, %
1	70	30
2	60	40
3	50	50

Pressing of the mixture was carried out on a PSU-125 hydraulic press. A single-sided pressing mold made of X12CrNiTi8-10 stainless steel was used to form the samples. Pressing was carried out in the pressure range of 10–20 MPa for 1 min. The samples had a cylindrical shape with a diameter of 20 mm and a height of 6 mm. The appearance of the hydraulic press is presented in Fig. 1.

Fig. 1. External view of the hydraulic press PSU-125.

Sintering of the samples was carried out in a modernized reactor for self-propagating high-temperature synthesis, developed by scientists of the Lutsk National Technical University [17] (Fig. 2).

Fig. 2. Modernized reactor for sintering in the SHS mode.

This reactor design allows obtaining samples with minimal energy consumption at the highest possible productivity. For maximum control over the SHS process, 2 thermocouples are installed in the reactor, above and below the test sample. This arrangement of thermocouples allows controlling the speed of combustion wave propagation and uniformity of sintering at all stages [18].

Sintered samples are presented on Fig. 3.

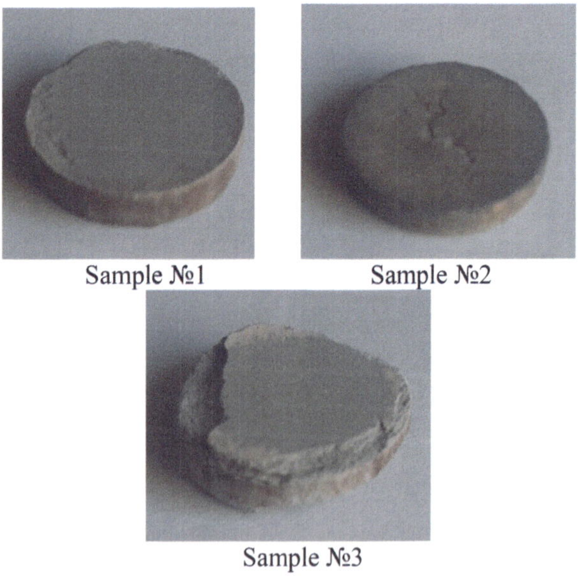

Fig. 3. Photo of samples (mixture from the table 1).

The porosity samples, which usually do not have through porosity, is measured by three main methods: hydrostatic weighing, mercury porometry and metallography [19, 20].

In this work, the porosity of the samples was investigated by hydrostatic weighing (Fig. 4). The results of the obtained data are true density, average density, open porosity and water absorption presented on Fig. 5.

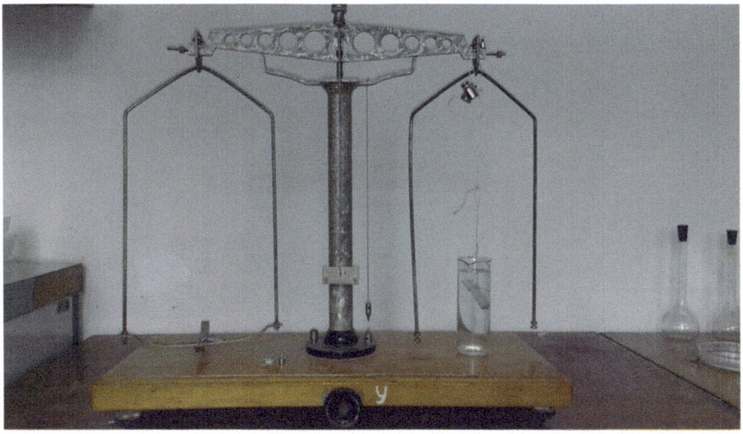

Fig. 4. Hydrostatic weighing.

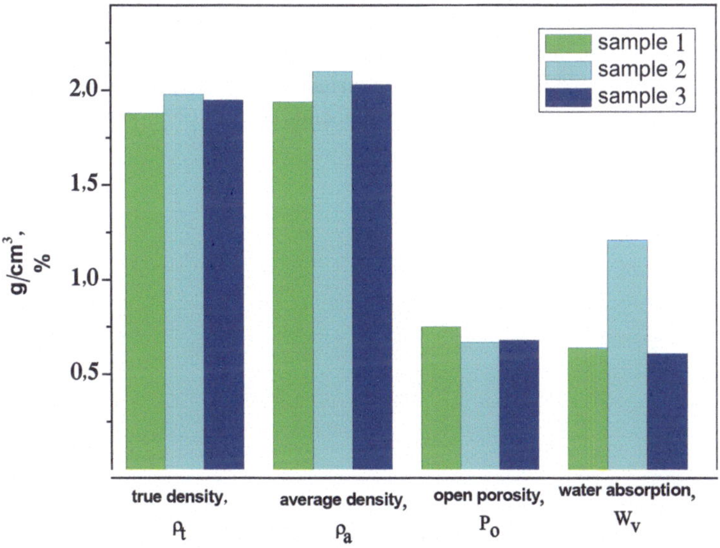

Fig. 5. Results of porosity and water absorption determination Sample№1 Sample №2, Sample №3 (diatomite + Ti).

4 Results and Discussion

The complex structure and high porosity of diatomite make it an excellent basis for modern permeable materials for various filtration.

SEM image of the porous material diatomite + Ti is presented in Fig. 6. From these images, it can be seen that the biocomposites diatomite + Ti indicate a heterogeneous porous structure with polygonal pores.

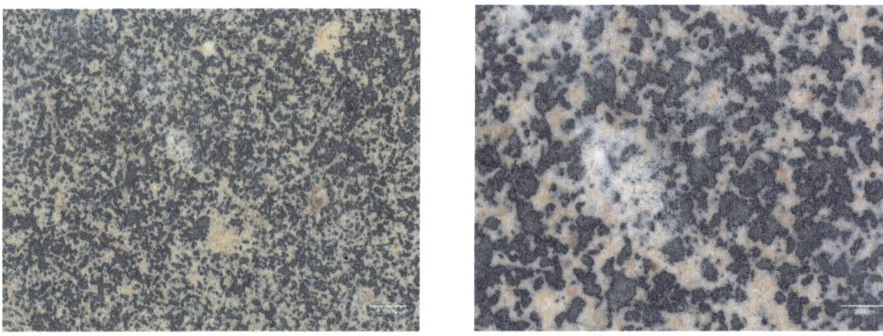

Fig. 6. SEM-image of porous material (diatomite + Ti).

To observe the differences in the structure of the porous material diatomite + Ti samples were subjected to profilographic analysis (Fig. 7). Measurements were performed for four parallel locations on the surface for each sample by method [21].

When creating a new type of filter material with the addition of diatomite, a structure is formed with an outer surface and an average roughness of Rz = 75 μm, Ra = 12.3 μm (Sample №2). With increasing pressure, the surface roughness decreases slightly (Fig. 8). The average roughness of the product also depends on the roughness of the stamp.

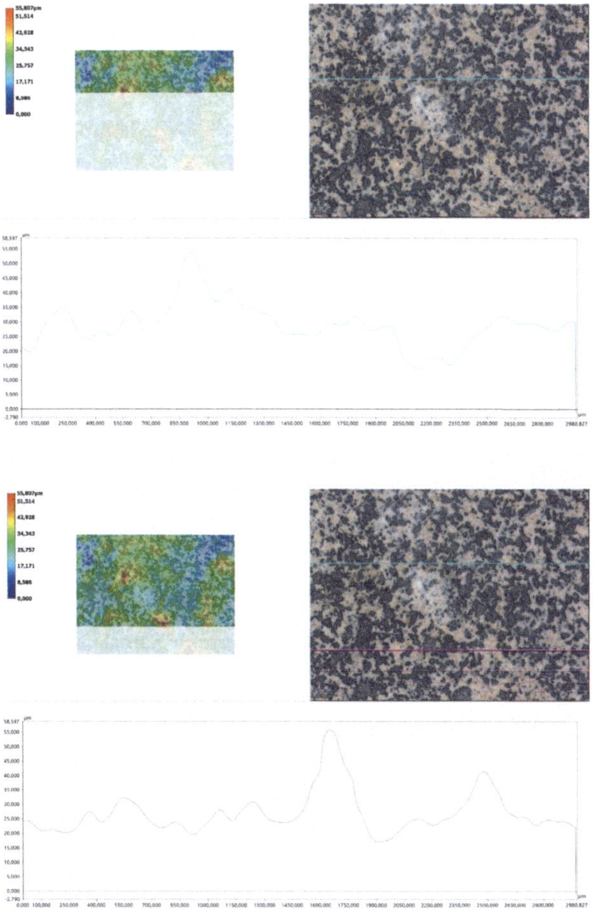

Fig. 7. Profile graph for each measured profile Sample №2 (diatomite + Ti).

Comparative studies of the surface quality of samples made of other materials were conducted, which are listed in Table 2.

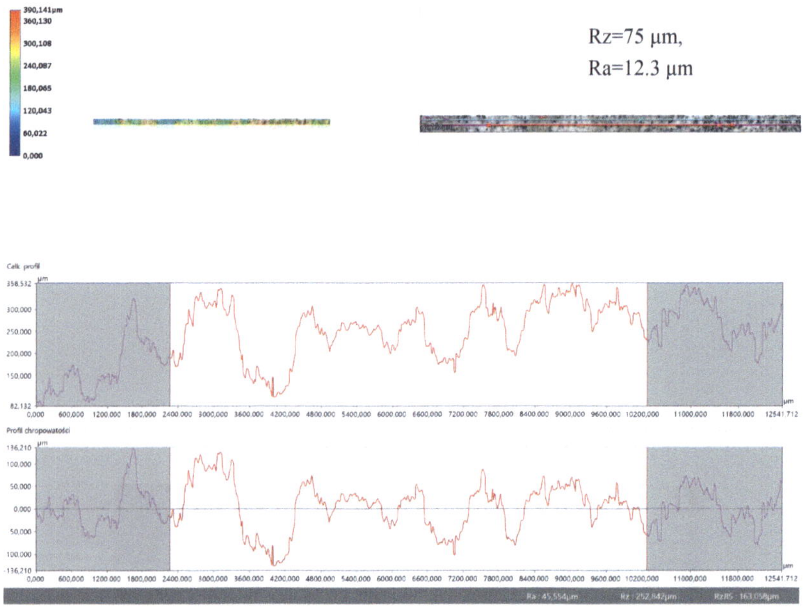

Fig. 8. Summarizing profilogram for the measured profile Sample №2 (diatomite + Ti).

Table 2. Surface quality of different filtration materials.

№	Materials	Ra, µm	Rz, µm
1	Diatomite + Ti	12.3…13.0	75.1…76.2
2	Ti	14.1…15.1	81.1…82.2
3	LH15	17.2…18.1	92.4…93.1

5 Conclusions

By investigating our materials, compressed diatomite + Ti, the article further highlights the versatility of Diatomaceous Earth in accommodating various materials, thereby expanding its application in future filtration technologies. An important aspect of the review is its emphasis on the environmental friendliness of Diatomaceous Earth. The environmentally friendly and biocompatible nature of Diatomaceous Earth distinguishes it from other materials, which often involve hazardous chemicals and complex processes.

The results obtained allow us to show the dependence of diatomite content on roughness. It was found that adding diatomite in a ratio of 60/40 to titanium allows us to create filter materials with a higher quality surface. Experimental studies have shown that when creating a new type of titanium-based filtering material with the addition of diatomite, the quality of the outer surface with an average roughness of Rz = 75 µm, Ra = 12.3 µm is formed.

In the future, comprehensive research will be conducted to create full-fledged diatomite filters with a developed surface for better and more productive work. Filters based on diatomite + Ti in further scientific research can be used to filter various types of liquids to eliminate pollutants, including various types of bacteria. If the filter is properly designed in the appropriate proportions, it can be used to remove bacteria up to 100%.

Acknowledgement. This work was carried out within the framework of a project BIOG-NET within the framework of the EU competition. The authors were direct participants and performers of certain scientific works from 16.01.2023 to 30.06.2023. The project is was implemented within the framework of the programme of the Foundation for Polish Science.

The authors express their sincere gratitude for the opportunity to participate in this project professor Krzysztof Jan Kurzydłowski (Bialystok University of Technology, Poland).

References

1. Kuhn, M., Briesen, H.: Dynamic modeling of filter-aid filtration including surface-and depth-filtration effects. Chem. Eng. Technol. **39**(3), 425–434 (2016). https://doi.org/10.1002/ceat.201500347
2. Anil, U.E., Al, E., Kayaci, K., Kara, F.: Fabricating of diatomite based ceramic water filter by a novel casting method. Adv. Sci. Technol. **91**, 48–53 (2014). https://doi.org/10.4028/www.scientific.net/AST.91.48
3. Miller, S.A., John, V.M., Pacca, S.A., Horvath, A.: Carbon dioxide reduction potential in the global cement industry by 2050. Cem. Concr. Res. **114**, 115–124 (2018). https://doi.org/10.1016/j.cemconres.2017.08.026
4. Li, K.: Ceramic Membranes for Separation and Reaction (2007). ISBN: 9780470014400
5. Povstyanoy, O., Zabolotnyi, O., Rud, V., Kuzmov, A., Herasymchuk, H.: Modeling of processes for creation new porous permeable materials with adjustable properties. In: Ivanov V., et al. (eds.) Advances in Design, Simulation and Manufacturing II. DSMIE 2019. Lecture Notes in Mechanical Engineering. Springer, Cham (2019). https://doi.org/10.1007/978-3-030-22365-6_46
6. Bruno, G., Efremov, A.M., Levandovskyi, A.N.: Connecting the macro- and microstrain responses in technical porous ceramics: modeling and experimental validations. J. Mater. Sci. **46**(1), 161–173 (2020). https://doi.org/10.1007/s10853-010-4899-0
7. Studart, A.R., Gonzenbach, U.T., Tervoort, E., Gauckler, L.J.: Processing routes to macroporous ceramics: a review. J. Am. Ceramic Soc. **89**(6), 1771–1789 (2006). https://doi.org/10.1111/j.1551-2916.2006.01044.x
8. Kokunešoski, M., et al.: Preparation and characterization of clay-based porous ceramics with boric acid as additive. Ceramics Int. **40**(9), 14191–14196 (2014). https://doi.org/10.1016/j.ceramint.2014.06.007
9. Sikora, P., El-Khayatt, A.M., Saudi, H.A., et al.: Rheological, mechanical, microstructural and radiation shielding properties of cement pastes containing magnetite (Fe_3O_4) nanoparticles. Int. J. Concr. Struct. Mater. **17**, 7 (2023). https://doi.org/10.1186/s40069-022-00568-y
10. Ghazanlou, S.I., Jalaly, M., Sadeghzadeh, S., et al.: A comparative study on the mechanical, physical and morphological properties of cement-micro/nanoFe_3O_4 composite. Sci. Rep. **10**, 2859 (2020). https://doi.org/10.1038/s41598-020-59846-y
11. Amin, M.S., El-Gamal, S.M.A., Hashem, F.S.: Effect of addition of nano-magnetite on the hydration characteristics of hardened Portland cement and high slag cement pastes. J. Therm. Anal. Calorim. **112**, 1253–1259 (2013). https://doi.org/10.1007/s10973-012-2663-1

12. Wu, C., Li, Z., Li, Y., Wu, J., Zhao, Y., Liao, Y.: Effect of starch on pore structure and thermal conductivity of diatomite-based porous ceramics. Ceramics Int. **49**(1), 383–391 (2023). https://doi.org/10.1016/j.ceramint.2022.08.352
13. Degirmenci, N., Yilmaz, A.: Use of diatomite as partial replacement for Portland cement in cement mortars. Constr. Build. Mater. **23**(1), 284–288 (2009). https://doi.org/10.1016/j.conbuildmat.2007.12.008
14. Łępicka, M., Rodziewicz, M., Kawalec, M., Nowicka, K., Tsybrii, Y., Kurzydłowski, K.J.: Diatomaceous earth as a drug-loaded carrier in a glass-ionomer cement. J. Mech. Behav. Biomed. Mater. **133**, 105324 (2022). https://doi.org/10.1016/j.jmbbm.2022.105324
15. Hasan, M., Saidi, T., Afifuddin, M.: Mechanical properties and absorption of lightweight concrete using lightweight aggregate from diatomaceous earth. Constr. Build. Mater. **277**, 122324 (2021). https://doi.org/10.1016/j.conbuildmat.2021.122324
16. Korzhyk, V., Berdnikova, O., Stukhliak, P., et al.: Structural features of the Cr_3C_2–NiCr and Ni–Cr–Fe–B–Si coatings produced by multichamber detonation spraying. Powder Metall. Met. Ceram. **63**, 107–116 (2024). https://doi.org/10.1007/s11106-024-00442-4
17. Samchuk, L., Rud, V., Povstyana, Yu., Saviuk, I.: SHS Processes in Industrial Waste Disposal Technologies, p. 150. LNTU, Lutsk (2017)
18. Rud V., et al.: Patent of Ukraine №91287 PMK (2014.01) B2F3/23 C01G1/00. Reactor for self-propagating high-temperature synthesis (SHS process), no. 12 (2014)
19. Dobrotvor, I.G., Stukhlyak, P.D., Mykytyshyn, A.G., Stukhlyak, D.P.: Influence of thickness and dispersed impurities on residual stresses in epoxy composite coatings. Strength Mater. **53**(2), 283–290 (2021). https://doi.org/10.1007/s11223-021-00287-x
20. Povstyanoy, O., Imbirovich, N., Tkachuk, V., Redko, R., Priadko, O.: Theoretical and experimental studies of the properties of porous permeable materials obtained from industrial waste. In: Ivanov, V., Trojanowska, J., Pavlenko, I., Rauch, E., Peraković, D. (eds.) Advances in Design, Simulation and Manufacturing V. DSMIE 2022, pp. 519–528. Springer International Publishing, Cham (2022). https://doi.org/10.1007/978-3-031-06025-0_51
21. Ryazantsev, A., Nechaev, V., Rebrova, S., Vorontsov, B., Bondarenko, O., Protasov, R.: Evaluation of the quality parameters of the surface layer of parts after machining with plasma heating. Strojnícky časopis – J. Mech. Eng. **74**(1), 123–134 (2024). https://doi.org/10.2478/scjme-2024-0013

Cellulose, Paper, and Textile Materials

Application of Eucalyptus Wood Cellulose in the Composition of Absorbent Types of Paper

Liudmyla Andriievska, Volodymyr Komakha(✉), Nina Merezhko, Nina Kaluha, and Nataliia Marchuk

State University of Trade and Economics, 19б Kyoto St., Kyiv 02156, Ukraine
{l.andriievska,v.komakha}@knute.edu.ua

Abstract. In this work, we investigated the use of eucalyptus cellulose in the composition of paper pulp to produce absorbent types of paper. Softwood cellulose was ground separately to a grinding degree of 29–36 SHR, and cellulose from eucalyptus to 17–23 SHR, respectively. The content of eucalyptus cellulose and softwood bleached sulfate cellulose in the paper composition is, respectively, 42–48 and 52–58 wt.%. The obtained macroporous structure of the developed paper provides high looseness, which is an indirect characteristic of important indicators of absorbent and filter types of paper, namely: air permeability, absorption capacity, and the presence in the composition of long fibers of sulphate bleached pulp and softwood at a high enough level. That is, the addition of cellulose from eucalyptus with a low degree of grinding allows to obtain paper with a uniform and closed structure, increased porosity, looseness and absorbency and to ensure at a high level its mechanical strength. This technological solution can also be used in the production of filtering paper.

Keywords: Pulp and Paper Industry · Eucalyptus Pulp · Paper Composition · Paper Structure · Degree of Grinding

1 Introduction

The pulp and paper industry is an important component of the economy of each country, which development is an indicator of economic growth and social development in general [1]. That is why scientists in this field should pay sufficient attention to the latest developments and innovations in paper production. In order for the products of domestic manufacturers to be competitive in relation to imported goods, modern raw materials, technologies and developments should be used in its production, which ensure an increase in the consumer properties of finished products [2].

One of these solutions is the use of alternative types of raw materials in the composition of paper. For a long time, the world pulp and paper industry for the production of paper products was dominated by northern tree species with soft wood, especially conifers, the use of which made it possible to produce high-quality long-fiber cellulose, which until recently was out of competition in the production of cardboard and paper of

F. J. G. Silva et al. (Eds.): DSMIE 2025, LNME, pp. 101–109, 2025.
https://doi.org/10.1007/978-3-031-96413-8_10

various types [3]. However, recently, cellulose from hard rocks, in particular eucalyptus, has become increasingly used [4].

Along with high quality characteristics, eucalyptus wood is much cheaper compared to coniferous wood species, and also has the properties of rapid restoration of plantations [5]. Eucalyptus belongs to fast-growing plants, the maturity period of which is limited to 5–7 years, it easily adapts to any type of soil and climatic conditions [6]. That is why eucalyptus trees in the future can become the main raw material source of the pulp and paper industry. The study of this issue has become relevant in the production of absorbent paper for sanitary and hygienic products and various types of filter paper [7].

The purpose of this work is to study the properties of eucalyptus cellulose as an alternative fibrous semi-finished product for the production absorbent types of paper.

2 Literature Review

Leading scientists of the world pulp and paper industry devoted their works to the study of the use of eucalyptus as an alternative raw material in paper production [8].

Brazilian scientists have investigated the process of forming the properties of cellulose for paper production depending on the age and species (genotype) of eucalyptus wood and the method of its processing [9]. Eucalyptus wood, which is the main raw material for pulp production in Brazil, has been found to be the most suitable for pulp production at the age of 5 years [10]. However, the authors of these scientific paper did not focus on the possibility of using eucalyptus cellulose in the production of absorbent paper, but provided only general recommendations on the production and possibility of using eucalyptus wood pulp in the medical, food, automotive industries, as well as for the production of viscose fabric.

A group of Portuguese scientists investigated the influence of the composition and physicochemical properties of bleached pulp from eucalyptus on its wettability and water-retaining ability [11]. It has been determined that production conditions that favor coarser and more deformed fibers (twisted and twisted) have an advantage for the production of pulp with a loose structure with good absorbency.

Another group of researchers from Portugal investigated the potential of unbleached pulp of spherical eucalyptus (the most common type of eucalyptus in Portugal) and Scots pine in terms of recyclability [12]. The structural, chemical and mechanical properties of both fibers were compared and found that eucalyptus fibers can withstand at least five times as many recycling cycles without losing their original characteristics [4]. Later Cunha and Simoes studied the effect of mechanical, chemical and enzymatic processing of spherical eucalyptus cellulose on its reactivity in the process of regeneration was studied [13]. The results obtained make it possible to assume that not only the original cellulose, but also the regenerated cellulose, has significant potential in the production of absorbent paper. Other scientists have also come to similar conclusions, in particular regarding the use of eucalyptus cellulose for evaporative cooling [14], heavy metal and pollutants biosorbent [15, 16], absorbent paper [17, 18].

Scientists from different countries also note the good mechanical and paper forming properties [19] of eucalyptus cellulose, which allows it to be used as a raw material for the production of nonwovens [20], cigarette paper [21].

However, the issue of using eucalyptus cellulose in the manufacture of paper for sanitary and hygienic products remains insufficiently studied. It is worth noting that the disadvantage of absorbent paper is a low level of strength with little exposure to moisture and low elasticity, which is undesirable for products of this type intended for wiping and suction of liquids. As soon as they are saturated with liquid, they become unsuitable for further use. Therefore, the task of creating a paper canvas that would maintain its integrity under the influence of moisture and behave like a sponge became important. This can be achieved by choosing compositions based on softwood cellulose fibers and eucalyptus wood cellulose, grinding methods and appropriate preparation of cellulose fibers to form the necessary structure and properties of paper [22].

Taking into account the fact that today there is no pulp production in Ukraine [23], and prices for finished products depend mainly on the cost of imported raw materials [24], it is relevant to increase the volume of eucalyptus pulp in paper production.

Due to the specific morphological structure and structure of fibers, eucalyptus wood cellulose has a high ability to form a homogeneous paper sheet and achieve correspondingly high indicators of strength, friability and whiteness. Therefore, the use of this pulp should contribute to the production of a paper web with increased capillary absorption capacity, which is an important consumer and operational property of this paper and products made of it under appropriate conditions of use.

To ensure increased indicators of mechanical strength and absorbency, we selected fibrous semi-finished products and established their ratio in the composition of paper pulp, namely the amount of long-fiber fraction used to create a structural framework and the amount of short-fiber fraction used as a filler or binder.

Thus, the purpose of this work is to achieve increased absorbency and looseness of paper at the required level of mechanical strength due to the use of alternative raw materials.

3 Research Methodology

For the study, paper samples were made according to five compositions, which differed in the ratio of fibrous components and their degree of grinding.

Paper samples were made as follows: sulfate bleached cellulose from coniferous wood was ground to a grinding degree of 29–36 SHR, and sulfate bleached cellulose from eucalyptus wood was milled to a grinding degree of 17–23 SHR.

The grinding operation is performed using laboratory pulsating rolls of the Valley mill at a rotation speed of 500 rpm in the presence of water.

Ground cellulose fibers to varying degrees of grinding were mixed with each other in water, obtaining a composition in different ratios by weights, %. The components are mixed in the composition regulators until they are homogeneous: mixer diameter - 75 mm, rotation speed - 250 rpm.

The process of obtaining paper included the following stages:

- grinding, preparation and production of two types of paper pulp from sulfate bleached pulp from coniferous wood and sulphate bleached pulp from eucalyptus wood;

- obtaining a composition of fibrous paper pulp based on a mixture of sulfate bleached pulp from softwood and sulfate bleached pulp from eucalyptus wood at their appropriate ratio;
- obtaining a fibrous suspension of paper pulp based on a mixture of the resulting composition;
- pouring a paper canvas from this suspension.

Experimental paper samples were produced using a Rapid-Kothen sheet-pouring laboratory device with a combined drying chamber (93 ± 1 °C drying temperature) according to ISO 5269-2:2004 and tested for mechanical strength, capillary absorption, friability and whiteness.

To compare the achieved level of properties in the obtained paper samples, an analogue sample (sample 6) was chosen, namely paper for sanitary and hygienic purposes. The paper pulp according to the analogue was obtained as follows: bleached sulfate cellulose from coniferous wood at a mass ratio of 50% was pre-ground to 20°SHR, followed by grinding together with sulfate deciduous cellulose and sulfite pulp from coniferous wood at a ratio of 30 and 20% according to the degree of grinding of 24°SHR.

Sampling for research was carried out in accordance with DSTU EN ISO 186:2008. Determination of the mass of 1 m^2 of paper was carried out in accordance with DSTU 2297-93.

Capillary absorption was determined by the Klemm method using a B-2T device ISO 8787:1986 Paper and board – Determination of capillary rise – Klemm method.

The indicator of the destructive force was determined on a RMB-30-2M vertical tearing machine according to DSTU 2334-94. The essence of the method is to determine the force that leads to the destruction of the sample during loading at a constant speed.

Friability was meant as a value wrapped in the thickness of the paper. Before determining destructive force, the weight per square meter (g/m^2), thickness in microns, and density (g/cm^3) are measured on 10 paper samples. The thickness of the castings is determined by the thickness indicator at a pressure of 0.1 MPa.

The whiteness of paper materials was determined according to DSTU ISO 2470:2005. The method is based on measuring the coefficient of diffuse reflection by the surface of the paper material of light in the blue region of the spectrum at an effective wavelength of light of 457 nm. Whiteness was measured on a spectrophotometer Konica Minolta CM-3630.

4 Results and Discussion

The results of the study and analysis of some morphological features of the structure and characteristics of eucalyptus cellulose in comparison with cellulose from hardwood of aspen, birch and beech are given in Table 1.

According to the above research results, aspen cellulose fibers are similar in length to eucalyptus cellulose fibers, but they have twice the width, and therefore have low flexibility. Eucalyptus fibers are shorter than aspen, birch and beech fibers and much thinner than them, the linear density of the fiber is only 60% of the indicator of birch fibers and 40% of beech fibers, which indicates that 1 g of eucalyptus cellulose contains twice as many fibers as birch cellulose and three times more than beech cellulose.

Table 1. Comparison of cellulose fiber properties fibers from aspen, birch, beech and eucalyptus wood

Indicator	Aspen, birch	Beech	Eucalyptus
Fiber length, mm	0.780–0.800	0.900–0.910	0.890–0.917
Fiber width, mm	0.022–0.025	0.014–0.018	0.011–0.013
Ratio of fiber length to width	35–32	64–51	81–71
Linear fiber density, mg/100m	8.0–8.2	12.5–12.8	4.6–5.6
Number of fibers in 1 mg	7950–8200	6200–6400	16200–19200

Sulfate bleached cellulose from coniferous wood and sulfate bleached cellulose from eucalyptus wood were used, based on the following considerations: sulfate cellulose fibers are more flexible, shorten to a lesser extent during grinding, fibrillating at the same time, which contributes to the production of a closed paper web. The combination of long and strong softwood cellulose fibers in a composition with shorter and wider eucalyptus wood cellulose fibers makes it possible to increase the absorbency of paper and its mechanical strength.

The main purpose of grinding short-fiber cellulose from eucalyptus wood for the production of absorbent paper is to increase the plasticity of fibers for development and provide it with strength properties. Therefore, we milled these types of pulp under mild conditions, that is, with minor additives of the surfaces of the grinding set.

The structure of absorbent or filter material, and especially of a low mass of 1 m^2, is a set of cellulose fibers randomly oriented in space, the axes of which are mainly located parallel to the plane of formation of a paper canvas from a fibrous aqueous suspension on the mesh of a paper machine. In this case, strength is provided only by the direct connection of cellulose fibers with each other.

The final degree of grinding was selected depending on the properties of the cellulose and the ability to retain the absorption properties during grinding at the appropriate mass with an area of 1 m^2 made from it. With separate grinding, coniferous pulp was ground to a higher degree of grinding of 29–36°SHR, while pulp from eucalyptus was ground to 17–23°SHR. The content of eucalyptus cellulose and coniferous bleached sulfate cellulose is 42–48 and 52–58 wt, %, respectively (Table 2).

The selected conditions for grinding pulp from coniferous and eucalyptus wood provide such conditions under which the fibers of each of the components are subjected to external and internal fibrillation. That is, the appearance of thin fibrils on their surface, which resemble thin hairs that contribute to the growth of the outer surface of the fibers and, as a rule, the number of points of contact, creating positive conditions for the occurrence of hydrogen bonds between the fibers during the formation of a paper canvas.

Separate grinding of sulfate bleached pulp from coniferous wood and separately sulphate bleached pulp from eucalyptus wood is also proposed, which makes it possible to preserve the length of the fibers as much as possible, which is important for obtaining a macroporous structure and provides a high level of absorbency, mechanical strength and friability of paper made from these fibers.

Table 2. Composition of the paper samples

Sample	Sulphate bleached softwood pulp, % wt	Sulfate bleached eucalyptus pulp, % wt	Sulphate hardwood pulp, % wt	Sulfite softwood pulp, % wt
1	52 (36°SHR)	48 (17°SHR)	–	–
2	54 (34°SHR)	46 (19°SHR)	–	–
3	56 (32°SHR)	44 (21°SHR)	–	–
4	58 (29°SHR)	42 (23°SHR)	–	–
5	60 (29°SHR)	40 (23°SHR)	–	–
6	50 (20 + °SHR)	–	30 (24°SHR)	20 (24°SHR)

Paper samples weighing 1 m^2 - 32 g were made from the obtained paper pulp compositions on Rapid-Kothen automatic sheet forming machine according to the scheme shown in Fig. 1.

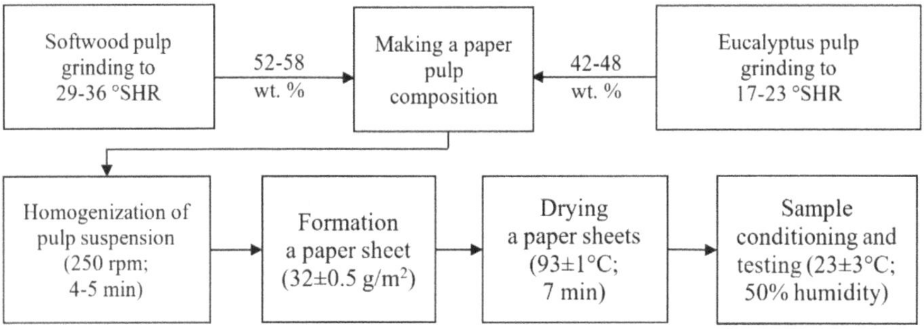

Fig. 1. Process of obtaining paper samples scheme.

Paper samples were tested in terms of density, mechanical strength (destructive force in the machine and transverse directions), capillary absorption and friability. Also, the whiteness index was controlled, which determines the aesthetic properties of finished paper and products made from it. The mechanical strength of paper was assessed by the indicator of destructive force, the suction capacity was assessed by the indicator of capillary suction on average in two directions for 10 min, looseness was determined as the inverse of the value of the density of paper. The results of the studies of paper samples are given in Table 3.

As evidenced by the results of the studies performed, the obtained paper samples are characterized by a high level of mechanical strength from 4.3 to 5.8 N, absorbency (from 27 to 36 mm in two directions), friability from 3.93 to 4.03 cm^3/g, which is ensured, in our opinion, by the optimal choice of a fibrous composition made of cellulose fibers of sulfate bleached pulp from coniferous wood and sulphate bleached cellulose from eucalyptus wood, their size and grinding conditions, which contributes to the formation

Table 3. Properties of the paper samples

Sample	Mass of 1 m^2, g	Density, g/cm^3	Destructive effort in machine direction, N	Destructive effort in transverse direction, N	Capillary absorption on average from two directions, mm	Friability, cm^3/g	Whiteness, %
1	31.9	0.248	4.3	2.2	27	4.03	86
2	32.1	0.249	4.6	2.3	34	4.00	86
3	32.1	0.250	4.9	2.5	36	4.00	85
4	32.3	0.254	5.8	2.9	38	3.93	85
5	32.4	0.254	5.8	2.8	36	3.93	85
6	32.4	0.337	3.9	1.8	24	2.96	84

of the necessary porous structure of the paper made from them. At the same time, it should be noted that the use of sulfate bleached pulp from softwood more than 58% does not lead to a significant improvement in the quality indicators of the developed paper, leads to an increase in its cost, and, accordingly, is inappropriate.

Microstructure of the developed paper (sample 4) and analogue (sample 6), which is shown in Fig. 2.

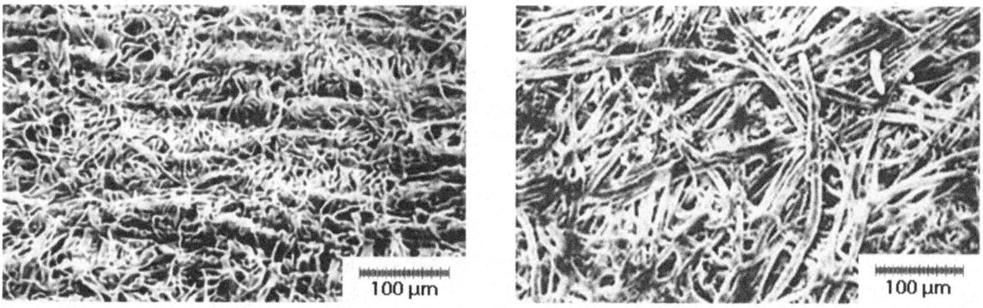

a) b)

Fig. 2. Microstructure: a) developed paper (sample 4); b) analogue paper (sample 6).

The obtained porous structure of the developed paper ensures the achievement of high friability, which is an indirect characteristic of important indicators of absorbent and filter types of paper, namely: air permeability, absorbency, and the presence of long fibers of sulfate bleached pulp from coniferous wood in the composition makes it possible to increase and ensure the mechanical strength of paper at a sufficiently high level. That is, the introduction of eucalyptus cellulose at a low degree of grinding (without a significant shortening of the size of the fibers) allows you to obtain paper with a uniform and closed structure, increased porosity, friability and absorbency and ensure its mechanical strength at a high level.

5 Conclusions

Paper produced as a traditional analogue, which incorporates a mixture of sulfate and sulfite cellulose from coniferous and hardwoods (birch, beech, aspen), exhibits a 10.2–48.7% lower level of mechanical strength compared to the developed paper. It is stiffer and less elastic, as evidenced by a 32.7–36.1 reduction in looseness (softness) and 12.5–58.3% in absorbency, which are crucial for absorbent types of paper and their products.

The best results of the studied characteristics of the developed paper are achieved with a low content of softwood cellulose fibers (56–58% wt.), which have a higher cost compared to eucalyptus cellulose, and, accordingly, this will contribute to reducing the cost of absorbent paper with a high complex of such properties as capillary absorption (27–36 mm), mechanical strength (4.3–5.8 N), friability (above 4 cm^3/g) and whiteness (85–86%).

Further investigation into the application of eucalyptus wood cellulose in absorbent paper could focus on optimization of alternative hybrid fiber compositions with nanocellulose integration and carboxymethylcellulose modifications of eucalyptus fibers to increase water retention and reduce fiber hornification.

References

1. Shang, D., Lu, H., Liu, C., Wang, D., Diao, G.: Evaluating the green development level of global paper industry from 2000–2030 based on a market-extended LCA model. J. Clean. Prod. **380**, 135108 (2022). https://doi.org/10.1016/j.jclepro.2022.135108
2. Del Rio, D., et al.: Decarbonizing the pulp and paper industry: a critical and systematic review of sociotechnical developments and policy options. Renew. Sustain. Energy Rev. **167**, 112706 (2022). https://doi.org/10.1016/j.rser.2022.112706
3. Chauhan, S., Meena, B.L.: Introduction to pulp and paper industry: Global scenario. Phys. Sci. Rev. **6**(5), 81–109 (2021). https://doi.org/10.1515/psr-2020-0014
4. Sembiring, N., Herlim, C., Febrilliandika, B.: Replanting Eucalyptus as a sustainable solution for pulp industry: a review. In: IOP Conference Series: Earth and Environmental Science, p. 012031. IOP Publishing (2024). https://doi.org/10.1088/1755-1315/1352/1/012031
5. Schubert, M., Panzarasa, G., Burgert, I.: Sustainability in wood products: a new perspective for handling natural diversity. Chem. Rev. **123**(5), 1889–1924 (2022). https://doi.org/10.1021/acs.chemrev.2c00360
6. Bayle, G.: Ecological and social impacts of eucalyptus tree plantation on the environment. J. Biodiver. Conserv. Biores. Manag. **5**(1), 93–104 (2019). https://doi.org/10.3329/jbcbm.v5i1.42189
7. Andrievska, L., Glushkova, T, Koptiukh, L.: Eucalyptus pulp as alternative raw material in paper production. Int. Sci.-Pract. J. Commod. Mark. **19**(1), 142–148 (2015)
8. Campos, J.F., Berteina-Raboin, S.: Eucalyptol, an all-purpose product. Catalysts **12**(1), 48 (2022). https://doi.org/10.3390/catal12010048
9. Dos Santos, V.B., da Conceição, G.J., Rodrigues, P.D., Cirilo, N.R.M., Almeida, M.N.F., Rocha, M.F.V., et al.: Impact of age and genotype on the quality of eucalyptus wood for dissolving pulp. Ind. Crops Prod. **218**, 118930 (2024). https://doi.org/10.1016/j.indcrop.2024.118930

10. Susaeta, A., Rossato, F.G.: Efficiency of pulp and paper industry in the production of pulp and bioelectricity in Brazil. Forest Policy Econ. **128**, 102484 (2021). https://doi.org/10.1016/j.forpol.2021.102484
11. Azevedo, C.A., Rebola, S.M.C., Domingues, E.M., Figueiredo, F.M.L., Evtuguin, D.V.: Relationship between surface properties and fiber network parameters of eucalyptus kraft pulps and their absorption capacity. Surfaces **3**(3), 265–281 (2020). https://doi.org/10.3390/surfaces3030020
12. Simões, R.M.: Recycling performance of softwood and hardwood unbleached kraft pulps for packaging papers. Tappi J. **22**, 73–86 (2023). https://doi.org/10.32964/TJ22.2.73
13. Cunha, A.E., Simões, R.M.: Effect of eucalyptus globulus pulp properties on fock reactivity. Cellulose, 1–18 (2024). https://doi.org/10.1007/s10570-024-06215-x
14. Doğramacı, P.A., Riffat, S., Gan, G., Aydın, D.: Experimental study of the potential of eucalyptus fibres for evaporative cooling. Renew. Energy **131**, 250–260 (2019). https://doi.org/10.1016/j.renene.2018.07.005
15. Gebretsadik, H., Gebrekidan, A., Demlie, L.: Removal of heavy metals from aqueous solutions using Eucalyptus Camaldulensis: an alternate low cost adsorbent. Cogent Chem. **6**(1), 1720892 (2020). https://doi.org/10.1080/23312009.2020.1720892
16. Alshiteewi, S.: Preparation of biosorbent from Eucalyptus trees and its application for removal of some pollutants in water. AlQalam J. Med. Appl. Sci. **1**(6), 255–259 (2023). https://doi.org/10.5281/zenodo.7976134
17. Cremonez, V.G., Bonfatti, E.A., Andrade, A., Silva, E., Klitzke, R.J., Klock, U.: Wood basic density effect of Eucalyptus grandis in the paper making. Matéria (Rio de Janeiro) **24**(3), e12420 (2019). https://doi.org/10.1590/s1517-707620190003.0735
18. Amorim, E.P., et al.: Wood evaluation of Eucalyptus pellita F. Muell. and Eucalyptus tereticornis Smith as potential for pulp and paper production. Revista do Instituto Florestal **33**(2), 139–149 (2021). https://doi.org/10.24278/2178-5031.202133202
19. Osyka, V., Koptyukh, L., Komaha, V., Shulga, O., Mostyka, K.: Paper-forming properties of cellulose of different types and degrees of grinder. Techn. Sci. Technol. **1**(15), 227–234 (2021). https://doi.org/10.25140/2411-5363-2019-1(15)-227-234
20. Rebola, S.M., Ferreira, J., Evtuguin, D.V.: Potential of bleached eucalyptus kraft pulp for applications in nonwoven fibrous fabrics. J. Eng. Fibers Fabr. **15**(11), 155892502098014 (2020). https://doi.org/10.1177/1558925020980146
21. Morais, F.P., Bértolo, R.A., Curto, J.M., Amaral, M.E., Carta, A.M., Evtyugin, D.V.: Comparative characterization of eucalyptus fibers and softwood fibers for tissue papers applications. Mater. Lett. X **4**, 100028 (2019). https://doi.org/10.1016/j.mlblux.2019.100028
22. Zhao, Y., Moser, C., Lindström, M., Henriksson, G., Li, J.: Cellulose nanofibers (CNF) from softwood, hardwood and tunicate: preparation-structure-film performance interrelation. ACS Appl. Mater. Interfaces. **9**(15), 13508–13519 (2017). https://doi.org/10.1021/acsami.7b01738
23. Nepal, P., Lamica, A., Parajuli, R.: Projected effects of the Russian invasion of Ukraine on global forest products markets. Forest Policy Econ. **168**, 103301 (2024). https://doi.org/10.1016/j.forpol.2024.103301
24. Kaba, G.: Review of the properties, acceptance, and use of eucalyptus as an alternative species in Ethiopia's wood industries. Indon. J. Innov. Appl. Sci. (IJIAS) **4**(2), 188–197 (2024). https://doi.org/10.47540/ijias.v4i2.1366

Study of the Carding Process and its Impact on the Strength of Spunlace Non-woven Fabric

Michał Sąsiadek[1](✉) and Milan Edl[2]

[1] University of Zielona Góra, 9, Licealna Street, 65-417, Zielona Góra, Poland
m.sasiadek@iim.uz.zgora.pl
[2] University of West Bohemia, 2732/8, Univerzitní Street, 30100 Pilsen, Czech Republic

Abstract. The article presents the results of the nonwoven fabric production process, specifically Spunlace, with particular emphasis on the carding process of the fibre blend and the final strength of the resulting nonwoven fabric. In the production test, conducted on an actual production line, two different types of carding machines were used to produce nonwoven fabric under identical production parameters. The nonwoven produced was subjected to microscopic examinations of its structure in terms of fibre bonding methods and strength tests. It was observed that differences in the design of the carding machines tested affect the final strength of the nonwoven fabric. The tested carding machines differed mainly in the arrangement of the drums between the two main cylinders, which is discussed in this article.

Keywords: Spunlace nonwoven fabric · Carding process · Strength

1 Introduction

The modern, global textile industry requires continuous development and improvement to remain competitive by reducing or maintaining current production costs while ensuring the expected and acceptable quality of the manufactured products.

Many manufacturers focus on meeting the needs of more demanding customers, competing on quality rather than just the price of their products. This highlights the importance of various studies aimed at optimising, improving, or introducing innovations in the field of nonwoven fabric production. The nonwoven production analysed in this article is carried out sequentially, starting with the mixing of appropriate fibre proportions, followed by the carding process of the blend, its further consolidation through hydro-entanglement and final drying to achieve the required end properties [1].

One of the most important processes is fibre carding, which is carried out on a machine called a carding machine. On this machine, fibres are separated, mixed, cleaned, and aligned using drums covered with appropriately configured carding cloth, which move in various arrangements, preparing the fibres for the stage of mutual entanglement [2].

F. J. G. Silva et al. (Eds.): DSMIE 2025, LNME, pp. 110–117, 2025.
https://doi.org/10.1007/978-3-031-96413-8_11

2 Literature Review

The literature reveals many areas related to the production and study of nonwoven fabric. These include, first and foremost: modelling and optimisation of processes, analysis of the microstructure and its impact on the properties of nonwoven fabric, testing the functional properties of nonwoven fabric and others.

In modelling and optimisation, much attention has been devoted to the modelling of the carding process in a system with a single drum and a fibre collection and return system [1–5], where phenomena occurring between the co-operating drums and the mixing of fibres, introduced into the system, are described. The studies also address the mathematical modelling of the relationships occurring in the carding process, including the development of a carding process model for a two-drum carding machine with an intermediate drum between the main cylinders [6, 7]. In turn, the article [3] models the operation of a two-drum carding machine with a system of four intermediate drums between the main cylinders. In articles [6–8], the dynamics of nonwoven fabric flow formation were studied in various configurations of two-drum carding machines, taking into account the uniformity of the density of the formed web and the fibre mass balance. These issues have a significant impact on the formation of the fibre blend, and consequently, its strength.

The subject of the work [9, 10] was the study of tensile strength and tear resistance of the nonwoven fabric produced through the carding process, including analysis of fibre orientation and bonding strength. Another aspect of the nonwoven fabric study was determining the impact of the carding process settings on its final strength, resulting from the mutual orientation of the fibres [10].

Many studies focus on analysis of the structure of the nonwoven fabric formed. These analyses are primarily conducted using scanning electron microscopy. These studies, among other things, focus on determining the mutual orientation of fibres and the spatial representation of the fibre blend [11–13], as well as the phenomenon of fibre deformation and changes in fibre orientation [14], where relationships between changes in fibre orientation, orientation angle and the degree of material deformation were also defined.

This article focusses on presenting the carding process of fibre blends from the perspective of its various methods, considering the design of two different devices and their impact on the final strength of the nonwoven fabric.

3 Research Methodology

The primary task of the carding process is to loosen, clean and arrange the fibres in parallel, as well as to preliminarily form the so-called fibre web. This is achieved by passing the raw material, in the form of a specific fibre blend, between the differently moving surfaces of drums covered with carding cloth. This process breaks up disorganised fibre clusters and then arranges the individual fibres in such a way that they are positioned parallel to one another.

In the nonwoven fabric production process studied, two different types of carding machines were used, arranged in a sequential configuration. This means that each machine can operate independently or simultaneously, even though they are set in a single line. A general image of the carding machine is shown in Fig. 1.

Fig. 1. Image of the carding machine construction in the process studied.

The carding machines used in the study differ in the method of transferring fibres from drum C1 to drum C2. In carding machine no. 1, the transfer of fibres between drums C1 and C2 is achieved through a single intermediate drum R1, while in carding machine no. 2, the transfer is done through a parallel arrangement of drums R1 and R2, as well as R3 and R4, as schematically shown in Fig. 2. The flow of carded fibres in the drawings is marked in red.

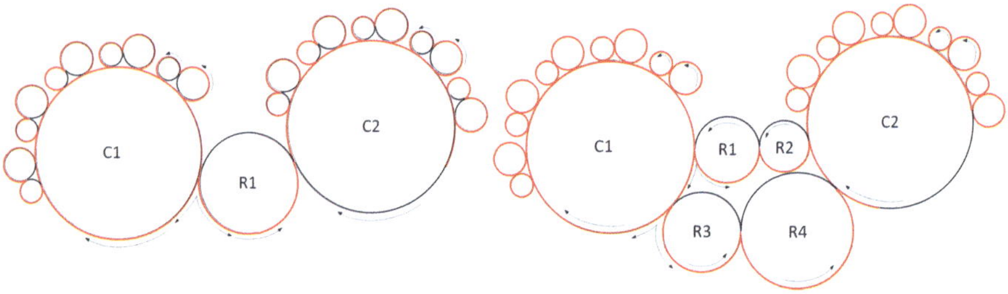

Fig. 2. Diagram of the main drum arrangement C1 and C2, as well as the intermediate transition between them, in carding machine No. 1 (on the left side) and carding machine No. 2 (on the right side).

3.1 Experimental Production of Nonwoven Fabric

Production of the nonwoven fabric was planned on an actual production line, in two experimental tests. Production of nonwoven fabric with the same surface weight, i.e., 40 gsm, made from a blend of polyester (80%) and viscose (20%), was adopted in these tests, with a process line speed of 190 m/min.

The first test (T1) involved the production of nonwoven fabric using carding machine No. 1, while the second test (T2) used carding machine no. 2.

After conducting technological trials for the production of the aforementioned nonwoven fabric, samples were taken for tensile strength testing and microscopic image analysis. These studies focussed on finding correlations between longitudinal (MD) and transverse (CD) tensile strength and the formation of different fibre bonding methods. An additional aspect of the study was the comparison of fibre mixing quality and orientation achieved in different industrial carding machines.

3.2 Sampling of the Nonwoven Fabric

After conducting the nonwoven fabric production tests and allowing the process to stabilise, samples were taken for microscopic and tensile strength testing. The samples, in the form of nonwoven fabric sections measuring 3.2 m in width and 1.5 m in length, were cut at five consecutive locations, approximately spaced at about 100 m, for the two experimental tests conducted.

3.3 Analysis of the Nonwoven Microstructure

For the microscopic analysis of the nonwoven fabric, 15 digital images were prepared from each of the 5 strips taken for each of the two tests. The images of the nonwoven fabric were taken using the JEOL JSM 7600F scanning electron microscope, which allows observation of the topography of a given organic or inorganic material in the range from nanometric to micrometric scale. All images of the nonwoven fabric tested were taken at approximately $\times$ 150 magnification with a scale of 100 μm, allowing the individual fibres to be observed.

3.4 Tensile Strength Testing

Each sample taken was tested for tensile strength in two directions, namely longitudinal – in line with the production flow (MD) and transverse to it (CD). In each measurement, 7 proportionally distributed samples across the width were prepared from 5 strips of nonwoven fabric taken from each test, in accordance with the ISO 9073–3:2023 standard. The tensile strength tests were conducted on a Zwick Z050 machine.

4 Results and Discussion

4.1 Methods of Fibre Bonding

Analysing the microscopic images of polyester and viscose fibres in Spunlance-type nonwoven fabric, several types of recurring bonds and the inter-twinings between them were observed, which affect the strength properties of the entire material.

They are classified as:

- fibre entanglement points,
- self-crossing fibres,

- self-crossing fibres with other fibres,
- fibres twisting into a spiral,
- "U"-shaped entanglements,
- clusters of entangled fibres.

The first and simplest types of entanglements considered are the points where fibres hook onto each other; this occurs on the water needling machine as a result of the impact of the water jets. Example images of this type of bonding are shown in Fig. 3.

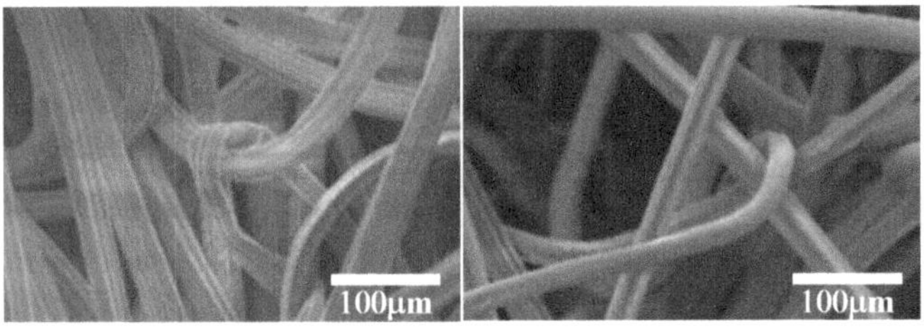

Fig. 3. An example of mutual fibre entanglement.

Another entanglement observed is the crossing of fibres, which can occur in the form of self-crossing or a series of multiple crossings. These situations are shown in Fig. 4.

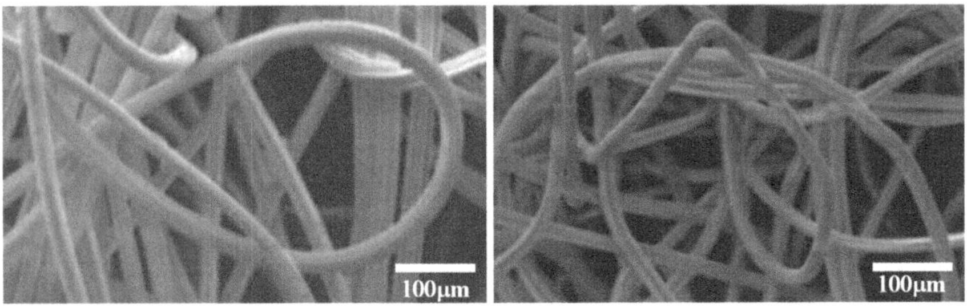

Fig. 4. An example of self-crossing fibres.

Another noticeable entanglement is the twisting of different fibres together, into a shape resembling a spiral. Examples of this type of entanglement from the samples tested are shown in Fig. 5.

Another type of entanglement is characterised by a more complex structure, as it consists of a large number of fibres bound together, forming a specific "U" shape. Examples of this type of entanglement, occurring in the nonwoven fabric, are shown in Fig. 6. It was also observed that, in most cases, the bonding is formed by fibres twisting into a spiral, which further affects the strength of the entanglement.

The last classified entanglement consists of all fibre clusters that intertwine, form knots, and bindings with shapes or proportions that are not precisely defined – Fig.7.

The arrangement of fibres during the carding process has a significant impact on the strength of the nonwoven fabric; however, the process that binds the fibres together is the

Fig. 5. An example of fibres entangled into a so-called spiral.

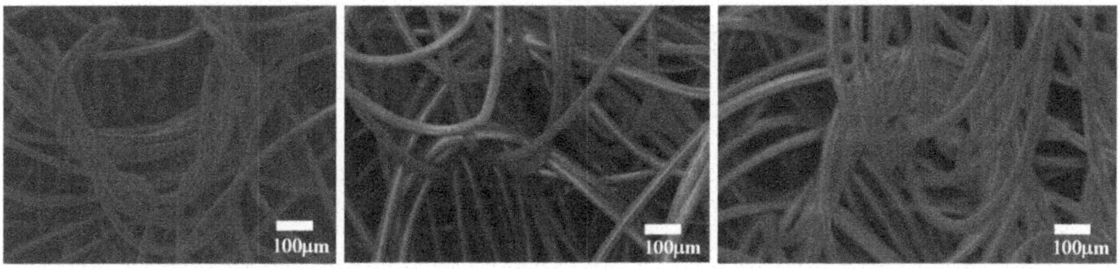

Fig. 6. An example of fibres entangled into a "U" shape.

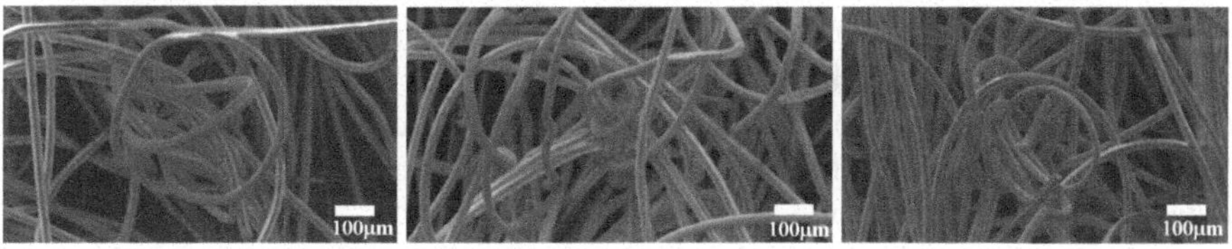

Fig. 7. An example of fibres entangled into so-called disordered clusters.

subsequent process after carding, namely the water needling process. In the tests conducted, this process was carried out identically for both tests, which allowed comparison of the tensile strength of the resulting nonwovens.

4.2 Tensile Strength of the Nonwoven Fabric.

Table 1 presents the results of tensile strength tests in the MD (Machine Direction) direction for samples taken from two examined production tests. The values MD1 to MD5 represent the average of 7 measurement samples in each case, while the value **MD** represents the average of the MD1 to MD5 values.

Table 2 presents, similarly to Table 1, the results of tensile strength measurements in the **CD** (Cross Direction).

Table 1. Values of the MD tensile strength measured.

Test number	MD1 [N]	MD2 [N]	MD3 [N]	MD4 [N]	MD5 [N]	**MD [N]**
Test 1 (card 1)	75.4	80.4	79.3	73.9	74.2	**76.64**
Test 2 (card 2)	81.9	71.2	81.7	79.2	80.1	**78.82**

Table 2. Values of the CD tensile strength measured.

Test number	CD1 [N]	CD2 [N]	CD3 [N]	CD4 [N]	CD5 [N]	**CD [N]**
Test 1 (card 1)	22.9	22.5	21.6	18.7	21.7	**21.48**
Test 2 (card 2)	24.5	23.3	23.5	22.8	22.6	**23.34**

5 Conclusions

In the production tests conducted, which form the research part described in this article, a nonwoven fabric was produced and subsequently subjected to microstructure analysis and tensile strength testing.

In the microscopic studies, aimed at identifying various fibre entanglements, six types of entanglements were characterised, which stand out in the microscope images analysed. In the nonwoven fabric produced in Test 2, using Carding Machine 2 (featuring a double roller arrangement between the main cylinders), a quantitatively higher number of entanglements was identified for each type characterised. This may affect the final strength of the nonwoven fabric.

In the tensile strength tests conducted in the MD and CD directions, higher strength was achieved in both directions for the nonwoven fabric produced in Test 2, as shown in Tables 1 and 2.

The higher strength of the nonwoven fabric produced using Carding Machine 2, in addition to the noticeably greater number of entanglements, may also result from the average time and distance travelled by the fibre during the carding process. Previous studies have shown that these parameters are higher in the process carried out on Carding Machine 2 compared to Carding Machine 1 [6–8], which is directly attributed to the design of the two carding machines.

The studies on the carding process described in this article do not fully address all the issues related to this process. The aim of subsequent research will be to analyse the strength of the nonwoven fabric based on tests considering different proportions of viscose and polyester fibre blends and varying weight loads of raw material fed into both carding machines.

Acknowledgment. The article was created as a result of a scientific internship and joint research at the University of West Bohemia in Pilsen, Czech Republic.

References

1. Albrecht, W.; Fuchs, H.; Kittelmann, W. (eds.): Nonwoven Fabrics: Raw Materials, Manufacture, Applications, Characteristics, Testing Processes. John Wiley & Sons, Hoboken (2006)
2. Batra, S. K.; Pourdeyhimi, B.: Introduction to Nonwovens Technology. DEStech Publications, Inc, (2012)
3. Karthik, T., et al.: Nonwovens: Process, Structure, Properties and Applications. WPI Publishing, (2017)
4. Russell, S.J. (ed.): Handbook of nonwovens. Woodhead Publishing (2006)
5. Ivars, J., Labanieh, A.R., Soulat, D.: Effect of the fibre orientation distribution on the mechanical and preforming behaviour of nonwoven preform made of recycled carbon fibres. Fibers **9**(12), 82 (2021)
6. Niedziela, M., Sąsiadek, M., Woźniak, W.: Modelling of the carding process for spunlace nonwovens with particular regard to selected mechanical parameters in a double-drum card. Part 1: Modelling of the fibre deck forming process. J. Text. Inst. **111**, 1–11 (2019)
7. Niedziela, M., Sąsiadek, M., Woźniak, W.: Modelling of the carding process for spunlace nonwovens with particular regard to selected mechanical parameters in a double-drum card. Part 2: Modelling of delay times in the longitudinal mixing process. J. Text. Inst. **11**, 1–10 (2019)
8. Niedziela, M., Sąsiadek, M., Woźniak, W.: Modelling of the selected mechanical properties of the modern double-drum cards for manufacturing of spunlace nonwovens. J. Text. Inst. **112**, 1655–1665 (2021)
9. Santos, A.S., Ferreira, P.J.T., Maloney, T.: Bio-based materials for nonwovens. Cellulose **28**(14), 8939–8969 (2021)
10. Roy, R., Ishtiaque, S.M.: Optimal design of a filter media by tuning the structure of needle punched nonwoven: influence of carding parameters. Fibers Polym. **21**(9), 2125–2137 (2020)
11. Hou, J., Ouyang, W., Xu, B., Wang, R.: Convolutional neural network for extracting 3D point clouds of fibrous web from multi-focus images. IEEE Access **8**, 87857–87869 (2020)
12. Chen, Y., Deng, N., Xin, B.J., Xing, W.Y., Zhang, Z.Y.: Structural characterization and measurement of nonwoven fabrics based on multi-focus image fusion. Measurement **141**, 356–363 (2019)
13. Kim, D.: Quantitative evaluation of geotextile void structures using digital image analysis. J. Kor. Geosynth. Soc. **12**(1), 51–61 (2013)
14. Stolyarov, O., Ershov, S.: Characterization of change in polypropylene spunbond nonwoven fabric fiber orientation during deformation based on image analysis and Fourier transforms. J. Strain Anal. Eng. Des. **52**(8), 457–466 (2017)

Engineering Education

The Use of Smart Complexes to Enhance Engineering Education and Student Effectiveness

Oleksandr Humennyi[1](✉), Serhii Kubitsky[2], and Yuriy Kubitskyy[3]

[1] Institute of Vocational Education of the National Academy of Pedagogical Sciences of Ukraine, 98a, Vito-Lithuanian Lane, Kyiv 03045, Ukraine
gumenniy7@gmail.com

[2] National University of Life and Environmental Sciences of Ukraine, 15, Heroyiv Oborony Street, Kyiv 03041, Ukraine

[3] Odesa State Agrarian University, 99, Kanatnaya Street, Odesa 65039, Ukraine

Abstract. The article explores the use of Smart complexes in educational subjects as an innovative approach to enhancing the effectiveness of training future professionals in engineering. It highlights the integration of advanced digital technologies into the educational process, focusing on their role in adapting learning materials to the individual needs of students.

The study results quantitatively confirmed the effectiveness of Smart complexes: students' creativity increased by 28%, critical thinking by 34%, and problem-solving skills by an average of 31%. The integration of self-assessment mechanisms ensured a 25% increase in student motivation and a 30% rise in their engagement in the learning process.

Thus, Smart complexes create a dynamic and personalized educational environment aimed at developing key competencies, preparing students for the challenges of the engineering profession in the digital age.

Keywords: Smart complex · digital technologies · professional training

1 Introduction

Today, numerous approaches to the digitalization of education exist globally, each of which deserves attention. For example, Bozkurt's [1] research analyzes hybrid and blended learning models, which enhance the personalization of the educational process through the integration of digital technologies. Other studies, such as Hrastinski [2], explore approaches to implementing blended learning in the context of educational innovations. Similarly, Shahaimi [3] highlights the importance of adapting digital platforms to ensure interactivity and student engagement in the learning process.

This perspective aligns with the proposed approach to the use of smart complexes, which also aim to integrate digital technologies to enhance interactivity, student engagement, and the personalization of the learning process.These complexes integrate modern

F. J. G. Silva et al. (Eds.): DSMIE 2025, LNME, pp. 121–135, 2025.
https://doi.org/10.1007/978-3-031-96413-8_12

digital technologies, enhancing the flexibility and personalization of the learning process, allowing students to engage more deeply with educational materials.

A Smart Complex for an academic subject is a comprehensive and multifunctional information system that integrates electronic educational resources in various formats (text, visual, audio, and interactive) into a unified information-educational environment. This system is designed to provide flexibility, interactivity, and continuity in the learning process. Smart complexes not only include electronic textbooks, interactive exercises, and multimedia presentations but also incorporate analytical tools for performance monitoring, self-assessment modules, adaptive learning systems, and collaboration tools such as chats, forums, and shared workspace features.It is developed for educational and methodological purposes and is designed to ensure a continuous and complete didactic cycle of the learning process. It is important to emphasize that smart complexes for academic disciplines represent an original approach that employs modern digital technologies to create individualized learning trajectories for students. The flexible structure of smart complexes enables the adaptation of the learning process to students' individual needs through interactive tasks, a modular learning system, the integration of digital tools, and real-time feedback.

According to the presented analysis, Smart Complexes outperform other solutions in most criteria, making them the optimal choice for modern education focused on integrating innovative technologies and fostering an individualized learning approach.

The constructive elements of Smart Complexes include environments such as creative, authorial, non-verbal, encyclopedic, self-realization, self-assessment, and information and communication technologies. These environments provide a clear structure for personalized learning, where creative, authorial, and self-assessment environments contribute to the development of critical thinking, creativity, and problem-solving skills. Such an approach aligns with innovative educational practices widely applied across European countries. At the same time, modern engineering education faces several challenges, including the rapid digitalization of the industry, the need to develop interdisciplinary competencies such as critical thinking, creativity, and digital literacy, as well as preparing students to work with modern technologies and tools. However, many traditional teaching methods remain insufficiently adapted to these challenges, creating a research gap in the development of innovative approaches to engineering education.

The modernization of engineering education is essential to align it with the needs of the digital economy, a globalized labor market, and new professional training standards. The use of Smart Complexes offers solutions to these challenges by integrating modern digital technologies, enabling the adaptation of educational materials to individual student needs, increasing their engagement in the learning process, and fostering key competencies necessary for professional activities in a rapidly changing world.

The developers of Smart Complexes for academic disciplines were instructors and methodologists from higher professional education institutions. They utilized Oleksandr Humennyi's proprietary methodology [4], which was applied to create Smart Complex projects for specific academic disciplines.

A comparison of Ukrainian Smart Complexes with European scientific research demonstrates that these domestic developments not only meet global standards but also actively implement innovative approaches to the individualization of the learning process (Table 1).

Table 1. Comparative Analysis of the Advantages of Smart Complexes Over Other Educational Solutions.

Criterion	Smart Complexes	Traditional Platforms	Mobile Applications
Personalized Learning	Interactive tasks, adaptive learning paths, and self-assessment modules ensure an individual approach	General materials without adaptation to individual needs	Provide personalization, but limited by app functionality
Interactivity and Engagement	Incorporates gamification, analytics, and interactive exercises	Basic forums and chats with limited interactive feedback	Includes interactive features, but lacks advanced options for deep interaction
Cognitive Skills Development	Promotes critical thinking, reflection, self-assessment, and problem-solving through metacognitive tools	Focused on knowledge transfer without emphasis on metacognitive development	Focuses on specific aspects like creativity or problem-solving but lacks a comprehensive approach
Accessibility and Scalability	Integrated with Microsoft Teams, accessible on any device, highly scalable	Requires specific software or limited to local servers	High accessibility, but dependent on specific platforms or resources
Progress Monitoring	Analytical modules enable real-time tracking of student performance	Basic statistics without in-depth data analysis	Offers limited monitoring capabilities, often restricted to specific tasks
Adoption of Modern Trends	Aligns with blended learning, gamification, and digital competency development	Supports only basic aspects of digital education	Focused on certain trends but lacks a comprehensive approach

The study by Pinto and Azevedo [5] emphasized the importance of aligning information and communication technologies (ICT) with pedagogical strategies to ensure effective learning, which became the foundation for the development of Smart Complexes. Their focus on adapting digital tools to the educational context and the needs of learners was reflected in the modular structure of Smart Complexes, which enables the personalization of the learning process.

The purpose of this article is to explore the implementation of Smart Complexes as an innovative tool to enhance the effectiveness of professional training in digital education, focusing on their ability to personalize educational materials according to student needs.

The main objectives of the article:

To clarify the concept of smart complexes and their application in the educational process for personalized learning.

To evaluate the impact of integrating digital technologies into academic subjects on the development of students' critical thinking, creativity, and analytical skills, as well as the effectiveness of smart complexes through interactive tasks, self-assessment modules, and visualizations aimed at enhancing student engagement.

To compare Ukrainian smart complexes with international educational approaches, particularly in blended learning, digital competencies, and gamification, to determine their relevance and alignment with global trends.

2 Literature Review

It is important to emphasize that Smart Complexes for academic disciplines are a regional Ukrainian development that leverages modern digital technologies to create individualized learning trajectories for students. These Smart Complexes were developed in pedagogical colleges in Berdychiv and Korostyshiv, as well as in vocational education institutions in Kharkiv, Kryvyi Rih, Khmelnytskyi, and Kropyvnytskyi.

The definition and implementation of Smart Complexes align with various areas actively explored in European research. For example, Bozkurt [6], in his systematic review of hybrid learning, highlights that using hybrid models facilitates the personalization of the learning process and the integration of digital technologies to achieve individual educational goals. This aligns with findings from the current research, where Smart Complexes enable task adaptation to students' preparation levels, creating an interactive learning environment for individualized learning trajectories.

Furthermore, Shahaimi's research [7] underscores the importance of integrating blended technologies into the learning process via digital platforms. This is directly related to the current study, as Smart Complexes utilize similar platforms to deliver interactive modules, enhancing student engagement and fostering competency development.

There is evidence that the integration of digital technologies into the learning process improves students' comprehension of complex technical materials [8], increases their engagement, and enhances learning outcomes [9]. Blended learning fosters more effective integration of digital tools with traditional teaching methods [2]. Smart Complexes adhere to these principles, offering an interactive environment for personalized learning trajectories and developing students' core competencies.

Rao and Desai [10] examined the impact of mobile applications on student engagement in higher education in India. Their research found that mobile apps provide flexible access to learning materials, and their strategic implementation significantly improves the quality of higher education. Similarly, Smart Complexes effectively address these challenges by deeply integrating digital technologies to ensure equal access to learning resources.

Studies on data mining technologies [11] have laid the foundation for the development of Smart Complexes. Identifying patterns in large datasets has enabled the adaptation of educational resources to meet learners' needs. Smart Complexes incorporate these technologies to individualize the learning process, facilitate effective interaction between learners and educators, and support personalized knowledge acquisition, aligning with European research trends.

Voelkel et al. [12] demonstrated that lecture capture technologies not only provide access to learning materials but also increase student engagement by allowing them to review content at their convenience. This approach was integrated into Smart Complexes by creating interactive modules with recorded sessions, ensuring personalized learning and fostering deeper knowledge acquisition.

Bruce and Kristiansen's work [13] highlighted that hybrid learning, combining traditional and digital methods, enhances flexibility, individualization, and learning efficiency. This approach has been implemented in Smart Complexes through the integration of blended educational formats, enabling the adaptation of learning to individual needs while maintaining access to digital resources and active interaction with educators.

Hartman [14] emphasized the importance of metacognitive skills in learning, highlighting the roles of self-reflection, planning, and monitoring in knowledge acquisition. These findings were integrated into the design of Smart Complexes through tools for self-assessment, learning planning, and feedback, contributing to the development of students' metacognitive abilities.

Numerous studies [15, 16] on assessing computational thinking through performance tests and self-assessment tools stress the importance of developing digital skills and understanding complex concepts via interactive, multi-level tasks [17]. These insights were applied to Smart Complexes by designing modules that foster the development of students' computational thinking.

The role of visual models in enhancing the understanding of complex concepts and stimulating critical thinking has been explored in various studies [18]. These ideas were incorporated into Smart Complexes through the integration of visualizations and simulations, facilitating deeper material comprehension and cognitive skill development.

3 Research Methodology

Smart Complexes for academic disciplines were developed on the Microsoft Teams platform, which is widely used for organizing the educational process and creating an interactive learning environment [19]. The development process included several stages of integrating environments aimed at ensuring effective learning.

To foster creativity, channels and tabs within Microsoft Teams were utilized to host interactive tasks, group projects, and individual assignments. The use of Whiteboard [20] facilitated collaboration among students, while Microsoft Forms and Polly [21] supported brainstorming sessions for idea generation.

Educators designed adaptive learning trajectories using Microsoft OneNote [22] and SharePoint [23], enabling the creation of individualized learning plans and the provision of personalized resources for students or groups. The integration of multimedia tools, such as PowerPoint, videos, graphic materials, and charts [24], ensured visual communication, helping students better grasp educational content.

Access to academic and reference sources was organized through integration with OneDrive, SharePoint, and external databases [25]. Educators created modern libraries of materials, including scientific articles, books, and other resources, which students could utilize for independent or group learning.

Microsoft Teams also facilitated the organization of individual and group projects, where students worked on research papers or creative assignments. Channels and specialized groups for projects provided conditions for collaboration, task distribution, and progress tracking using tools like Task Planner or Assignments [26]. This environment supported the realization of students' potential in project work, which is essential for their professional development.

Detailed criteria for assessing students' creativity and critical thinking during the use of Smart Complexes are presented in Table 2.

Smart Complexes utilized technological platforms to create a personalized learning environment and track student performance. Tools such as self-assessment tests, interactive tasks, and visual models fostered the development of creative thinking and critical analysis while simultaneously monitoring students' progress.

Data collection included surveys, observations of activities, analysis of digital products (models, videos), and questionnaires to evaluate student satisfaction and the integration of digital technologies.

The integration of Smart Complexes into Microsoft Teams enhanced student engagement, optimized the learning process, and provided a flexible assessment system. Indicators were developed to evaluate the effectiveness of self-assessment mechanisms, measuring students' awareness, engagement, and progress.

Each indicator was assessed on a scale from 0 to 100, ensuring accuracy and objectivity in evaluation. The table below demonstrates each indicator and its corresponding evaluation scale (Table 3).

Evaluation Scale:

- 0–20 – Very low level
- 21–40 – Low level
- 41–60 – Average level
- 61–80 – High level
- 81–100 – Very high level

To assess the impact of Smart complexes, an analysis of baseline and post-implementation indicators of self-assessment, engagement, and student progress was conducted. Following integration into Microsoft Teams, improvements in critical thinking, creativity, and independence were observed.

The use of interactive tasks and visual models facilitated personalized learning and problem-solving skills development. Statistical analysis (paired t-test, ANOVA) confirmed significant improvements in academic performance and key competencies.

Table 2. Criteria for assessing students' creativity and critical thinking in the use of Smart complexes.

Criteria	Levels				
	5 (Very High)	4 (High)	3 (Medium)	2 (Low)	1 (Very Low)
Generation of new ideas. The evaluation was conducted through project analysis, self-assessment, and group evaluation	The student generates absolutely new and innovative ideas that exceed expectations	Ideas are new and adequate	Ideas have been used before but are effective	The number of new ideas is limited or unoriginal	Finds it difficult to generate new ideas, uses standard approaches
Flexibility of thinking. The assessment was conducted using cognitive flexibility tests	Easily changes the approach, adapts to new conditions	Adapts well, but sometimes needs more time	Adapts to new tasks but prefers to use tested methods	Has difficulty adapting to new tasks or changes	Unable to change thinking or adapt to new tasks
Originality in problem-solving. Evaluation was conducted through the analysis of creative solutions to problem situations	Proposes unique solutions that other students do not consider	Proposes new solutions but based on well-known approaches	Uses known methods, but applies them effectively	Proposes banal, non-innovative solutions	Finds it difficult to come up with new solutions
Problem analysis. The analysis was conducted using case study methods	Analyzes the problem well, identifies key aspects and offers deep analysis	Analyzes well, but sometimes misses details	The analysis is shallow but helps solve the problem	Has difficulty analyzing problems, focuses on surface-level details	Unable to effectively analyze the problem
Evaluation of arguments. Evaluation of arguments was conducted using debate and peer review methods	Thoroughly evaluates arguments, considers all aspects, provides convincing evidence	Evaluates well, but sometimes misses less obvious aspects	The evaluation is basic, but the student proposes some deep analysis	Surface-level evaluation, struggles to differentiate between weak and strong arguments	Unable to evaluate arguments, provides weak evidence
Logical problem-solving. The evaluation was conducted through logical reasoning tests and the analysis of students' solutions to real-world tasks	Uses logical arguments, the solution is coherent	Uses logic, but the solution may lack structure	Logic is used but partially, often relies on intuition	The solution is based on incomplete or illogical reasoning	Uses irrational or illogical methods for problem-solving

Table 3. Indicators for Evaluating the Effectiveness of Student Self-Assessment Mechanisms through the Integration of Smart Complexes in Microsoft Teams.

Indicator	Description	Scale
Clarity of Understanding of Evaluation Criteria	Measures how well students understand the criteria by which they are expected to evaluate their tasks. Improved understanding is expected after implementing Smart complexes	0 (no understanding)—100 (full understanding)
Level of Self-Assessment Using Tools	Reflects the effectiveness of using self-assessment tools like Rubrics and Microsoft Forms. Higher scores indicate better use of tools for objective self-assessment	0 (tools not used)—100 (effective use)
Motivation for Self-Evaluation	Indicates the level of motivation students have for self-assessment after the introduction of new mechanisms	0 (no motivation)—100 (high motivation)
Monitoring Progress	Assesses students' ability to track their own progress using platform features such as Insights in Microsoft Teams	0 (no monitoring)—100 (full control over progress)
Engagement in the Self-Assessment Process	Measures the level of student engagement in the self-assessment process, including participation in reflection and evaluation of their results	0 (no engagement)—100 (high level of activity)
Reflection and Journal Keeping	Reflects how much students engage in self-reflection and journal-keeping through tools like OneNote	0 (no reflection)—100 (regular reflection)
Impact of Gamification on Self-Assessment	Assesses how elements of gamification, such as Kahoot! and Quizlet, influence students' self-assessment and engagement	0 (no gamification influence)—100 (strong positive influence)

4 Results and Discussion

The study confirmed the key role of Smart Complexes in enhancing education by fostering creativity, critical thinking, and problem-solving skills. The integration of these innovative technologies into the digital environment of Microsoft Teams significantly improved the effectiveness of future professionals' training, enhancing students' self-assessment and creative self-realization.

An analysis of quantitative and qualitative indicators revealed a substantial increase in cognitive skills, confirming the positive impact of Smart Complexes on the learning process. The comparative table (Table 4) illustrates the levels of these skills before and after the experiment.

The experiment confirmed that Smart Complexes align with modern educational trends, including blended and hybrid learning, digital competencies, and gamification. This approach ensures personalized learning and fosters key competencies essential for adaptation to digitalization and globalization.

Integrated into Microsoft Teams, Smart Complexes combined interactive tasks, self-assessment modules, visualization, and gamification, enhancing critical thinking, creativity, and problem-solving skills. The use of digital tools such as Microsoft Forms and analytical modules enabled the creation of an interactive learning environment with automated data collection and flexible learning trajectory adjustments.

Statistical analysis (t-tests, ANOVA) confirmed significant improvements in cognitive skills, including creativity, critical thinking, adaptability, and self-organization.

The success of this approach lay in creating a dynamic and interactive learning environment that ensured flexibility, personalization, and student engagement, enhancing the overall efficiency of the educational process. This organization of learning contributed not only to academic achievements but also to the development of essential skills for the modern digital world.

Thus, the study confirmed the feasibility of using Smart Complexes as a structured, adaptive, and student-centered approach that aligns with global trends in digital education. Through such innovations, students are better prepared for contemporary challenges, developing competencies that meet the demands of a digital and interconnected world.

All research objectives were successfully achieved, specifically confirming the effectiveness of Smart Complexes as an innovative tool for enhancing education quality, ensuring personalized learning, and developing key student competencies to help them adapt to the challenges of a globalized and digital society.

The experimental results indicate a significant improvement across all criteria for assessing students' creativity and critical thinking after the implementation of Smart Complexes. On average, scores for each indicator increased by 1.2 to 1.7 points, highlighting the positive impact of the technology on students' skill development.

Generation of new ideas. Before the experiment, the average score was 2.8 ± 0.3, indicating a moderate ability to generate new ideas. After the experiment, the score rose to 4.5 ± 0.2, reflecting a high level of innovative thinking and the ability to create original solutions.

Flexibility of thinking. The initial score before the experiment was 3.1 ± 0.4, which increased to 4.3 ± 0.3 post-experiment. This demonstrates improved adaptability to new conditions and tasks, facilitated by the personalized learning environment created by Smart Complexes.

Originality in problem-solving. The average score before technology implementation was 2.9 ± 0.5, indicating reliance on familiar approaches. Post-experiment, the score increased to 4.2 ± 0.3, reflecting enhanced creativity in solving complex problems.

Table 4. Cognitive Skills Improvement Before and After Experiment.

Criteria	Before Experiment (Average Score, scale 1–5)	Error (±, scale 1–5)	After Experiment (Average Score, scale 1–5))	Error (±, scale 1–5)
Generation of new ideas	2.8	±0.3	4.5	±0.2
Functionality of thinking	3.1	±0.4	4.3	±0.3
Originality of solutions	2.9	±0.5	4.2	±0.3
Analysis of problems	3.0	±0.4	4.4	±0.2
Evaluation of evidence and arguments	2.7	±0.3	4.1	±0.3
Logical problem solving	2.9	±0.4	4.2	±0.2

Problem analysis. Scores increased from 3.0 ± 0.4 to 4.4 ± 0.2, showing improved ability to analyze problems with greater attention to detail and depth.

Evaluation of evidence and arguments. Students initially scored relatively low, at 2.7 ± 0.3, but post-implementation, the score rose to 4.1 ± 0.3, demonstrating significant progress in evaluating arguments and evidence.

Logical problem-solving. The average score before the experiment was 2.9 ± 0.4, increasing to 4.2 ± 0.2 after the experiment, highlighting improved ability to think logically and systematically.

The margin of error in the research results was minimal, confirming the reliability of the data and the effectiveness of Smart Complexes in enhancing students' creativity and critical thinking.

The experiment demonstrated that implementing self-assessment mechanisms through Smart Complexes in Microsoft Teams improved academic performance, particularly in self-assessment, motivation, and engagement. The analysis of key indicators confirmed students' successful adaptation to new progress monitoring tools and reflective practices (Table 5).

Table 5. Results before and after implementing self-assessment mechanisms through the integration of Smart complexes in Microsoft Teams.

Indicator	Before Implementation (%)	After Implementation (%)	Change (Percentage Points)
Clarity of understanding assessment criteria	55	85	+ 30
Level of self-assessment using tools	50	82	+ 32
Motivation for independent evaluation	45	80	+ 35
Progress monitoring	52	88	+ 36
Engagement in the self-assessment process	60	90	+ 30
Reflection and journaling	40	75	+ 35
Impact of gamification on self-assessment	48	84	+ 36

The implementation of self-assessment mechanisms through the integration of Smart Complexes into the Microsoft Teams platform significantly improved the learning process outcomes. After implementation, the following positive changes were observed:

Clarity of understanding assessment criteria increased by 30 points, indicating better student comprehension of evaluation requirements and success indicators.

The level of self-assessment using tools increased by 32 points, demonstrating enhanced effectiveness in students' use of digital tools for evaluating their achievements.

Motivation for independent evaluation rose by 35 points, reflecting heightened interest and engagement of students in the self-assessment process.

Progress monitoring improved by 36 points due to the use of analytical tools that enable real-time tracking of student performance.

Engagement in the self-assessment process increased by 30 points, indicating higher student activity in the learning process.

Reflection and journaling improved by 35 points, showing enhanced student ability to analyze and evaluate their achievements.

The implementation of Smart Complexes significantly improved students' self-assessment (+36 points), motivation, engagement in the learning process, and ability to monitor their progress.

The experiment confirmed the positive impact of innovative educational technologies on cognitive skills, particularly critical thinking, creativity, problem analysis, and logical reasoning. The use of self-assessment mechanisms enhanced students' autonomy and learning efficiency.

Microsoft Teams and Microsoft Forms provided an interactive learning environment, automated result analysis, and instant feedback, contributing to personalized learning. Video conferencing, chats, and teamwork improved communication between educators and students.

Challenges included varying levels of digital competence among educators, limited resources in educational institutions, and the difficulty of quantitatively measuring skills such as creativity and critical thinking.

Overall, Smart Complexes proved effective in developing key competencies and improving learning outcomes, opening new opportunities for the digital transformation of education.

Research by Hartman [27] emphasizes that metacognition, which includes planning, monitoring, and evaluating one's actions, is a critical component of successful learning. Similarly, the results of our experiment demonstrate that the interactive tools of Smart Complexes, such as self-assessment modules, journaling, and gamification integration, created conditions conducive to developing these skills. For instance, students using Smart Complexes significantly improved their progress monitoring skills (+36%) and reflective abilities (+35%), confirming the effectiveness of this approach in fostering metacognitive competencies.

Smart Complexes have already been integrated into the educational processes of some Ukrainian HEIs, particularly for engineering training. Their features include both open-access modules and licensed resources. Basic functionality in Microsoft Teams is available for free through the Microsoft Education program, while specialized analytical and gamification modules may require additional licenses.

In the EU, Smart Complexes have strong potential for engineering education due to their adaptability, personalization, and alignment with educational reforms. They support the development of innovative thinking, adaptability, and interdisciplinary skills.

The demand for digital learning environments in blended and distance education creates opportunities for integrating Smart Complexes into European HEIs through international partnerships. Their implementation enhances educational outcomes and fosters active learning through interactive methods.

5 Conclusions

The study of the concept of Smart Complexes has confirmed their effectiveness in personalized learning, optimizing the educational process, and increasing its adaptability to the individual needs of students. The integration of digital technologies into the learning process contributed to the development of critical thinking, creativity, and analytical skills. Special attention was given to interactive tasks, self-assessment modules, and visualizations, which significantly enhanced student engagement in the learning process.

1. Improvement in Cognitive Skills:
   - Student creativity increased by **28%;**
   - Critical thinking improved by **34%;**

- Problem-solving ability grew by **31**%.

2. Enhancement of Motivation and Self-Assessment:

   - Student motivation rose by **25%;**
   - Engagement in the learning process increased by **30%;**
   - Skills in independent progress monitoring improved by **32%,** indicating significant growth in self-organization and autonomy.

3. Alignment with Global Standards:

   A comparison of Ukrainian Smart Complexes with international educational approaches in blended learning, digital competencies, and gamification confirmed their alignment with global educational trends.

4. Future Research Directions:

   - Evaluation of the long-term impact of Smart Complexes on graduates' professional skills in various fields;
   - Refinement of interactive elements for deeper personalization of learning;
   - Exploration of the effectiveness of Smart Complexes in distance and blended learning formats, considering varying levels of student preparedness.

Thus, the results of the study confirmed that Smart Complexes are an effective tool for enhancing the efficiency of the educational process. They not only address modern educational challenges but also contribute to the development of key competencies, preparing students for the demands of a digital and globalized world.

References

1. Bozkurt, A.: A retro perspective on blended/hybrid learning: systematic review mapping and visualization of the scholarly landscape. TechTrends (2022). https://doi.org/10.1007/s11528-022-00696-0
2. Hrastinski, S.: What do we mean by blended learning? TechTrends (2019). https://doi.org/10.1007/s11528-019-00375-5
3. Shahaimi, S.: Pengintegrasian Blended Learning dalam Pembelajaran Persekitaran Maya Frog. Academia.edu (2016). https://www.academia.edu/31671754/Pengintegrasian_Blended_Learning_dalam_Pembelajaran_Persekitaran_Maya_Frog
4. Humennyi, O., Radkevych, O., Radkevych, V.: An environmental approach to developing and applying smart complexes of academic disciplines in professional training of future specialists. Revista Românească pentru Educaţie Multidimensională **13**(2), 516–539 (2021). https://doi.org/10.18662/rrem/13.2/434
5. Pinto, A., Azevedo, J.: Integrating ICT into teaching: the Portuguese experience. Int. J. Educ. Dev. Using Inf. Commun. Technol. **7**(3), 32–43 (2011). https://doi.org/10.4018/978-1-4666-1918-0.ch010
6. Shahaimi, S.: Pengintegrasian Blended Learning dalam Pembelajaran Persekitaran Maya Frog. Academia.edu. https://www.academia.edu/31671754/Pengintegrasian_Blended_Learning_dalam_Pembelajaran_Persekitaran_Maya_Frog
7. Svyrydiuk, V., Luzan, P., Svyrydiuk, O., Titova, O., Popova, O.: Electronic textbooks as means for developing professional competencies of engineering students. In: Tonkonogyi, V., Ivanov, V., Trojanowska, J., Oborskyi, G., Pavlenko, I. (eds.) Advanced Manufacturing Processes III.

InterPartner 2021. Lecture Notes in Mechanical Engineering, pp. 68–76. Springer, Cham (2022). https://doi.org/10.1007/978-3-030-91327-4_68
8. Liu, Y., Han, S., Li, H.: Understanding the factors driving M-learning adoption: a literature review. Campus-Wide Inf. Syst. **27**(4), 210–226 (2010). https://doi.org/10.1108/10650741011073761
9. Titova, O., Luzan, P., Sosnytska, N., Kulieshov, S., Suprun, O.: Information and communication technology tools for enhancing engineering students' creativity. In: Tonkonogyi, V., Ivanov, V., Trojanowska, J., Oborskyi, G., Pavlenko, I. (eds.) Advances in Design, Simulation and Manufacturing IV. DSMIE 2021. Lecture Notes in Mechanical Engineering, pp. 332–340. Springer, Cham (2021). https://doi.org/10.1007/978-3-030-77719-7_33
10. Rao, P., Desai, V.: Impact of mobile learning applications on student engagement in indian higher education. Asian J. Distance Educ. **18**(3), 67–82 (2023). https://www.asianjde.org/ojs/index.php/AsianJDE/article/view/492
11. Triayudi, A., et al.: Improving the Quality and Education Systems Through Data Mining. ScienceDirect. https://www.sciencedirect.com/science/article/pii/S266699352100045X
12. Voelkel, S., et al.: Lecture Capture Affects Student Learning Behaviour. ScienceDirect. https://www.sciencedirect.com/science/article/pii/S0742051X22001454
13. Bruce, A., Kristiansen, R.K.: Hybrid futures for innovative learning. Comput. Educ. (2022). https://doi.org/10.1016/j.compedu.2022.104397
14. Hartman, H.J. (ed.) Metacognition in Learning and Instruction. Springer, Heidelberg (2002). https://doi.org/10.1007/978-94-017-2243-8
15. Titova, O., Luzan, P., Yarosh, L., Symonenko, S., Zaitseva, N.: The test instruments development based on the items complexity. In: 2022 IEEE 4th International Conference on Modern Electrical and Energy System (MEES), Kremenchuk, Ukraine, pp. 1–6. IEEE (2022). https://doi.org/10.1109/MEES58014.2022.10005697
16. Guggemos, J., Seufert, S.: Measuring computational thinking: adapting a performance test and a self-assessment instrument for German-speaking countries. In: 16th International Conference on Cognition and Exploratory Learning in Digital Age (CELDA 2019), pp. 63–70. IADIS Press (2019). https://doi.org/10.33965/celda2019_201911L009
17. Titova, O., Luzan, P., Davlatzoda, Q.Q., Mosia, I., Kabysh, M.: The taxonomy approach for engineering students' outcomes assessment. In: Tonkonogyi, V., Ivanov, V., Trojanowska, J., Oborskyi, G., Pavlenko, I. (eds.) Advanced Manufacturing Processes IV. InterPartner 2022. Lecture Notes in Mechanical Engineering. Springer, Cham (2023). https://doi.org/10.1007/978-3-031-16651-8_36
18. Spitzer, M.: The Mind within the Net: Models of Learning, Thinking, and Acting. MIT Press, Cambridge (1999). https://doi.org/10.7551/mitpress/1481.001.0001
19. Microsoft Teams: Features and Benefits for Education. https://www.microsoft.com/en/microsoft-teams
20. Whiteboard for Collaborative Learning. Microsoft. https://www.microsoft.com/whiteboard
21. Polly: Surveys and Polls for Teams. Microsoft. https://www.polly.ai/microsoft-teams
22. Microsoft OneNote for Education. Microsoft. https://www.microsoft.com/en-us/education/onenote
23. SharePoint in Education: Organizing Resources. Microsoft. https://www.microsoft.com/en-us/sharepoint
24. Multimedia Tools in Teams. Microsoft. https://www.microsoft.com/en-us/microsoft-365/powerpoint
25. Integration of OneDrive and External Databases in Education. https://www.microsoft.com/en-us/microsoft-365/onedrive
26. Task Planner and Assignments in Teams. https://www.microsoft.com/en-us/education/microsoft-teams

27. Hartman, H.J.: Developing Students' Metacognitive Knowledge and Skills. https://www.researchgate.net/publication/274710464_Developing_Students%27_Metacognitive_Knowledge_and_Skills

Methodical System for Development of the Professional Competence of Future Electrical Engineering Teachers: Experimental Verification of Implementation

Andrii Kalenskyi[1](✉), Oleksandra Borodiyenko[2], Liubov Pavliuk[3], Nataliia Vanina[1], and Inna Hrytsenok[1]

[1] Institute of Vocational Education of the National Academy of Educational Sciences of Ukraine, 98a, Vito-Lithuanian Lane, Kyiv 03045, Ukraine
kalenskyandrii@gmail.com

[2] Institute of Higher Education of the National Academy of Educational Sciences of Ukraine, 9, Bastionna Street, Kyiv 01014, Ukraine

[3] State Institution "Scientific and Methodological Center of Higher and Professional Pre-Higher Education", 11, Smilyanska Street, Kyiv 03151, Ukraine

Abstract. The results of the experimental verification of the implementation of the methodical system for development of the professional competence of future electrical engineering teachers are presented. The implementation of the methodical system was based on the regularities related to the internal structure of the system and its external connections. The components of the methodical system: content, methods, forms and technologies for future professionals training were. It was proved that the efficiency of the development of future teachers' professional competence in electrical engineering depended on the methodology, educational technologies and forms for students' learning and cognitive activities. It was substantiated that the important component in such a process was a scientifically substantiated and verified methodical system based on the model of professional competencies future electrical engineering teachers, the theoretical foundation of competence-based training and up-to-date scientific findings in pedagogy aiming for continuous improvement of their pedagogical and technological knowledge and skills, personal and professional qualities, development of general culture and value orientations. It was determined that the purposeful development of future teachers' professional competence in electrical engineering is possible through the use of pedagogical technologies, such as personality-oriented technology, project-based and problem-based learning, use of information and communication technologies. The effectiveness of the implementation of the methodical system for the development of the professional competence of future electrical engineering teachers at Ukrainian universities has been experimentally proven.

Keywords: methodical system · electrical engineering · professional competencies of future electrical engineering teachers · pedagogical technologies

F. J. G. Silva et al. (Eds.): DSMIE 2025, LNME, pp. 136–147, 2025.
https://doi.org/10.1007/978-3-031-96413-8_13

1 Introduction

The urgency of the research is connected with Ukraine's integration into the European educational and research space, ongoing vocational and professional education reform in Ukraine, increasing requirements for professional training of graduates, the need to move from the traditional academic style of teaching to innovative one.

Current trends in education are directly related to the market requirements for professionals at this stage of social-economic development. The Law of Ukraine "On Education" in particular emphasizes that the purpose of higher education is to acquire a person with a high level of scientific, professional and general competencies necessary for activities in a particular specialty or a particular field of activity [1].

Analysis of current trends in higher education in Ukraine shows a contradiction between the market's need to train creative, purposeful teachers capable of continuous self-improvement and pedagogical interaction, and the inconsistency of content, teaching methods and forms at universities and colleges that provide this training. In addition, the constant change of educational programs requires active implementation in the educational process of training personal development and competence paradigms of education, which determines the appropriate selection of content, methods, forms and technologies of teacher training, their active involvement in self-education, and experience of cooperation with participants. Teachers play a significant part in the process. They train future professionals in the area of material production. Thus, the design, implementation and experimental verification of the methodical system as an effective component of the development of future teachers' professional competence in electrical engineering which is determined by the rapid development of competency-oriented educational process is in actual need.

The purpose of the article is an experimental verification of the effectiveness of the implementation of the methodical system for development of the professional competence of future electrical engineering teachers in universities in Ukraine.

2 Literature Review

Vocational and technical educational institutions should train modern highly qualified specialists of the appropriate level and profile, qualified, erudite, responsible, socially and professionally mobile [2], who are fluent in their profession, possess knowledge of the basics of scientific labor organization, and are capable of highly productive work for the needs of society.

Therefore, the training of future teachers of electrical engineering in universities is a priority task. Given the rapid changes taking place in modern society in recent decades, the higher education system is undergoing significant structural changes and is characterized by several priorities and values. In this regard, there are new requirements for the process of pedagogical training of students who will be future teachers aimed at forming the personality of a well-trained expert, ready to be effective in the classroom, realize continuous self-improvement, interaction and cooperation with colleagues, employers, who are ready for the updated content of integrated education. In addition, there is a need for further implementation of distance [3], problem-based [4], work-integrated [5], cooperative [6] learning and dual education [7], which enables effective development of competencies. The UN Competency Development Guide describes competence in detail as a combination of skills, characteristics and behavior that are directly related to job success. Competence is important for all professionals, regardless of profession, function or level. Moreover, the list of proposed competencies of the specialist is used as the basis of the human resources development systems, including enrolment, employment and performance appraisal [8, 9]. The concept of competence approach requires effective student-centered two-way communication between teachers and students. Objective and reliable assessment of the engineering students' learning outcomes and professional competence is also essential [10–12].

Therefore, we can conclude that, despite significant theoretical and practical experience, there is a lack of a clear methodology or methodical system that would allow preparing future electrical engineering teachers for teaching in vocational education institutions that are designed to train modern highly qualified specialists in the field of electrical engineering.

3 Research Methodology

The purpose of the research was experimental verification of the methodical system for of the professional competence of future electrical engineering which has been substantiated, designed and implemented within the process of future technologies for teachers' training in HEIs in Ukraine. The algorithm of actions to confirm or refute the research hypothesis included: for a parallel type of experiment - comparison of control (CG) and experimental (EG) groups and establishing the coincidence of their initial characteristics; realization of pedagogical influence on EG; based on the comparison of CG and EG after the implementation of the pedagogical impact of establishing differences in their final characteristics. For the sequential type of experiment, the initial and final characteristics of one group (EG) were compared.

The methodical system of the development of the professional competence of future electrical engineering teachers was based on the system approach to the research of the phenomena; theories of vocational education and training and management of engineering students' educational and creative potential while they are being trained at university [14, 15].

The methodical system should have the following characteristics: integrity, componentity, structure, compactness, functionality, multilevel, controllability and connection with the environment [16].

The project-oriented approach in education contributes to the disclosure and development of the hidden potential of the individual and the generation of innovative mechanisms for the organization of his life activities to solve socially significant tasks and regulate social processes [17, 18].

The article [19] examines the peculiarities of the formation of the methodical system for the training of future technology teachers and the need for its implementation through a systematic combination of the content of training, methods of training, forms, means and results of training.

The methodical system of the development of the professional competence of future electrical engineering teachers includes such components (Fig. 1) as purposes, content of training and educational technologies (forms, methods and means of training). It has been developed as a set of structural and functional components interconnected based on purposefulness, adaptability, optimality and efficiency. The content of the methodical system includes the knowledge, skills and personal abilities that future teachers should possess.

Methods of training that help the development of the professional competence of future electrical engineering teachers are: visual (demonstration, illustration, followed by discussion and reflection, self-observation), practical (practical training, laboratory work, visual and research activities, project-based method) and verbal (lecture, storytelling, explanation, instruction, consultation, discussion and coaching conversations). Forms of development of professional competence in electrical engineering are extracurricular (consultations, clubs, competitions, conferences, workshops), classroom training, non-formal (research based on literature analysis, Internet surfing, conferences, attending exhibitions of modern equipment, internships, dual training). The purposeful development of the professional competence of future electrical engineering teachers is also possible through the use of pedagogical technologies, such as personality-oriented technology, project-based and problem-based learning and use of information and communication technologies. Among the means of development of the professional competence of future electrical engineering teachers are: verbal (educational and scientific literature, didactic materials), technical (sound, screen, movies and multimedia), visual (schemes, diagrams, photographs, tables, boards and graphs), information and educational environment (electronic textbook, manuals, STEAM-laboratories, service platform systems etc.).

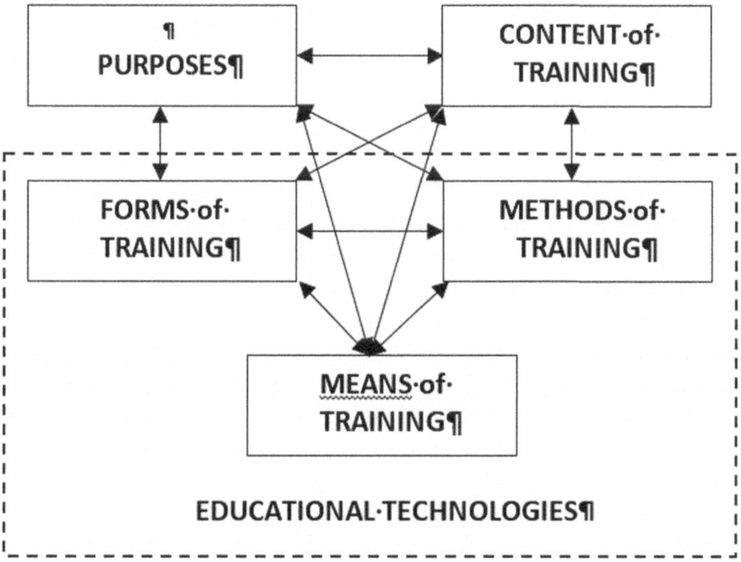

Fig. 1. Components of the methodical system of development of the professional competence of future electrical engineering teachers

In addition, in the methodical system of development of the professional competence of future electrical engineering teachers, a special role is played by the system of evaluation of learning outcomes, technology of enrolment and accumulation of educational credits for all possible options for development of individual trajectories of training. The criteria and indicators of levels of development of the professional competence of future electrical engineering teachers have been revealed: cognitive (knowledge of calculation of an electric circuit, electric cars of direct and alternating currents, etc.); activity (mastery of basic laws and ability to solve technical problems); personal (interest in technical activities as an important component of training, interest in increasing knowledge and skills in electrical engineering) and levels (initial, low, medium and high).

An electrical engineering lesson is a teaching form when the learning process is organized so, that the components of the lesson (the content, learning methods and organizational forms) exist in conjunction and are determined by the purpose of the lesson. The main purpose of the lesson is that students are fluent in the material presented. The process of achieving this goal is a consistent solution to such tasks as the acquisition of new knowledge and methods of action, independent exploration and development of professional values.

The purpose of the experiment, which was conducted for verification of the methodical system of development of the professional competence of future electrical engineering teachers was confirmation of the research hypothesis or its refutation and, accordingly, the relevance of theoretical results. The essence of the experiment was the purposeful influence of the methods of teaching and content to develop students' knowledge and skills in electrical engineering. The elements of such influence were the specific content (defined based on teachers' professional competence model and in accordance with the educational program), methods of training (by time, by number of students, by place of study), forms of training (by source of information, by main didactic task, by essence of cognitive activity, by level of integrity), means of training (verbal, technical, visual, ICT and educational environment).

To assess the progress in the development of components of the professional competence of future electrical engineering teachers, we used an ordinal scale of measurements. In particular, to measure the level of development of the value-motivational and subjective components of professional competence in electrical engineering, a 10-point scale was used, which reflects the level of propensity to a particular model; to measure the level of development of the intellectual-cognitive component of professional competence in electrical engineering - assessment on a relative scale of the results of mastering the relevant training courses (from 0 to 100%); to measure the level of development of the operational component of professional competence in electrical engineering - a scale from 1 to 5 points. We use methods to measure the levels of professional competence of the studied category of students in different ways (self-assessment, expert assessment, assessment of the level of knowledge acquisition) using different measurement scales (10-point scale that reflects the level of propensity to a particular model; the results of mastering the relevant training courses (from 0 to 100% for each completed task), assessment using the absolute scale of skills and abilities described by the indicator (from 1 to 5 points), it is advisable to convert to so-called aggregate indicators (in particular, indicators of distribution of the number of EG members according to the levels of development of professional competence in electrical engineering) were used to solve the problem of comparison of indicators before and after the pedagogical impact on EG.

The following algorithm was used to analyze the results of the measuring which showed the levels of the professional competence of future electrical engineering teachers development according to the main components: statistical processing of the obtained measurement results (descriptive statistics); evaluation of the level of the coincidence of the characteristics of CG and EG before the implementation of pedagogical influence; evaluation of the level of differences in the characteristics of CG and EG after the implementation of pedagogical influence.

To check the effectiveness of the substantiated methodical system, students were issued a questionnaire. Respondents were asked to evaluate the effectiveness of the methodical system, in particular, during the learning process. They were offered to solve the following tasks: use of forms, methods and teaching aids as a holistic process, finding out students' interest in implementing information obtained, use of service platforms, use of STEM laboratories in the educational process, use of electronic textbooks and electronic manuals; determine what skills they have after taking a virtual course; to search the level of development of competencies in electrical engineering. At the beginning of the experiment to determine the impact of the proposed technique, an experimental group was gathered. Students from another subgroup were stated as a control group.

At the statement stage of the experiment, a control assessment was conducted with the help of tests and a survey of students on the problem of identifying the effectiveness of the methodical system and modern techniques of training. Assessment of the professional competence of future electrical engineering teachers was carried out based on a three-component structure of criteria and relevant indicators: cognitive (knowledge of electric circuits, electric machines of direct and alternating currents); activity (possession of basic laws and ability to solve technical problems); personal (interest in technical activities as an important component of future training, interest in increasing knowledge and skills in the field of electrical engineering).

Four levels of the professional competence of future electrical engineering teachers were used: initial - the student has partial knowledge of the basic concepts in the study of electric and magnetic circuits, linear circuits of single-phase current, the main components of machines of permanent and alternating currents, aware about laws of transition from real to estimated electrical circuit; low - the student can distinguish a simple electrical circuit from a complex one, uses devices for measuring electrical and non-electrical quantities, has mastered the basic methods of calculating simple circuits, has some skills in constructing electrical circuits; medium - the student has the basic concepts of the basic laws of electrical engineering, aware about basic parameters of electrical circuits, performs simple calculations, is able to build volt-ampere characteristics of nonlinear elements; high – the student has knowledge of the principle of operation of transformers, DC and AC machines, low-voltage switching equipment, is able to make electrical circuits, explores the modes of electrical equipment, is able to build vector diagrams, has skills in calculating AC and DC circuits.

4 Results and Discussion

At the statement stage of the experiment, most students of higher educational institutions, future teachers were characterized by an initial level of the professional competence of future electrical engineering teachers. There was no significant difference in the development levels of the control and experimental groups, which was confirmed by experimental verification. Moreover, the levels of the professional competence of future electrical engineering teachers of the control group were slightly higher than those of the experimental group (Fig. 1). With the help of the criterion $\chi 2$ it is possible to test the hypothesis about the equality of probabilities of possessing the "initial", "low", "medium" and "high" levels according to certain indicators in the control and experimental groups of students. Data analysis showed that more than half of respondents in the observed groups had an initial and low level of professional competence in electrical engineering (within the control group it is 76.92%, within the experimental - 80.13%) (Table 1).

Table 1. Quantity of students (by levels of the professional competence of future electrical engineering teachers) (statement stage)

Group Status	Initial level	Low level	Medium level	High level	Total Quantity
Control group	77	33	28	5	143
Experimental group	89	32	27	3	151

Analysis of the results of the statement stage (Fig. 2) revealed that at the statement stage of the experiment, the level of professional competence in electrical engineering of future teachers within the experimental and control groups was mainly low.

As a result of calculations, it is established that criterion $\chi 2 = 5{,}96$. Thus, using the rules of decision-making using this criterion it is found that the experimental value of this criterion is less than the critical one, which confirms the null hypothesis about the similarity of the levels of the observed indicator in the studied groups.

During the period of the formative stage, the students of the experimental group were given interactive lectures and prepared for laboratory classes with the help of independent virtual training with subsequent implementation in the workplace. At the end of the experiment (after the final certification in the discipline "Electrical Engineering") a control assessment was conducted, which showed an improvement in the level of electrical knowledge in both control and experimental groups.

After the implementation of the developed methodical system, there were obvious changes in the development of the professional competence of future electrical engineering (Table 2).

The indicators in the experimental group are significantly higher than in the control group (Fig. 3).

After statistical processing, it was established that criterion $\chi 2 = 30{,}06$. Thus, using the rules of decision-making using this criterion it is found that the experimental value of this criterion is higher than the critical one, which confirms the hypothesis about the

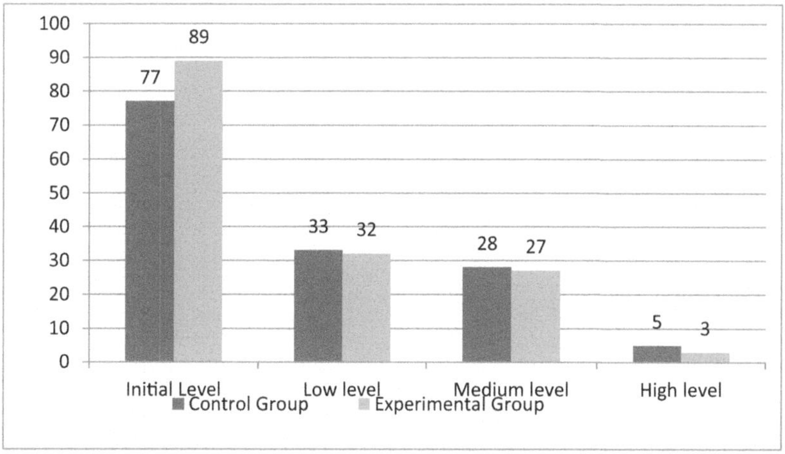

Fig. 2. Levels of development of the professional competence of future electrical engineering teachers (statement stage)

Table 2. Quantity of students (by levels of the professional competence of future electrical engineering teachers) (formative stage).

Group Status	Initial level	Low level	Medium level	High level	Total Quantity
Control group	19	36	56	32	143
Experimental group	11	29	63	48	151

difference of the levels of the studied indicator in the control and experimental groups. It should be emphasized that the professional competence of future electrical engineering teachers in the control groups was developed based on traditional teaching methods, and its development was carried out indirectly and unsystematically.

Most teachers who were involved in the training process with experimental groups, paid more attention to the development of the professional competence of future electrical engineering teachers of students in the process of pedagogical interaction, using information and educational environment of pedagogical activity based on the developed methodical system. As a result, higher indicators in experimental groups were obtained for assessment of the students' learning outcomes.

The obtained results partially confirm the conclusions of the scientist O. Markova that when creating a model of the methodical system it is necessary [20]: to take into account the professional focus of training future specialists; ensure the fundamentality of education by implementing a methodical system; predict the results of pedagogical influence by forming the professional competence of future specialists.

Despite the positive results of the experiment, one should take into account the objective and subjective factors identified by I. Kotsan, which are caused by social transformations, technical achievements, stereotyping and formalism of the existing system of professional training of teachers. Their low motivation and value orientations

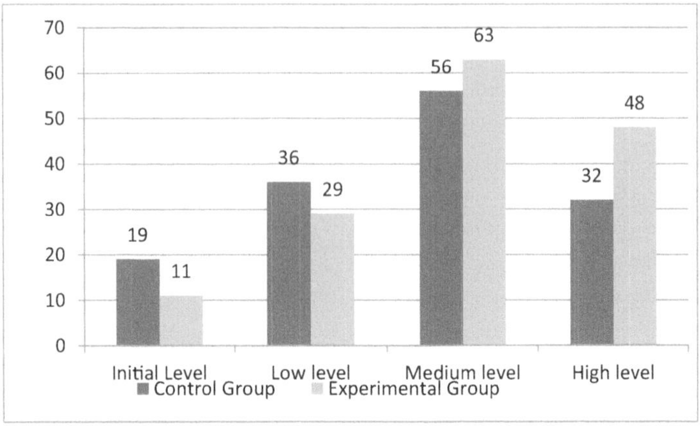

Fig. 3. Levels of development of the professional competence of future electrical engineering teachers (formative stage)

can become an obstacle to the implementation of a methodical system for development of the professional competence of future teachers of electrical engineering, in particular [21]:

the decline in the prestige of the teaching profession due to low social status, which is reflected in the amount of wages and the labor intensity of the teaching load;
insufficient level of computerization and material support of the educational process;
overloading students with mandatory educational tasks;
lack of modern textbooks and teaching-methodical aids and quality teaching-methodical materials for seminar, practical and laboratory classes, etc.

The directions of further research in the formation of the professional competence of future electrical engineering teachers of electrical engineering should take into account four new trends in engineering education, identified by R. Hudgraft and A. Kolmos [22]: the transition from teacher orientation to student orientation, active learning; more involvement in the practice of context engineering; use of digital technologies learning tools and personalized learning, proven portfolio.

5 Conclusions

The obtained results of the experiment and their statistical processing made it possible to develop several conclusions: the level of the professional competence of future electrical engineering teachers in the control and experimental groups were statistically equal; the respondents in the observed groups in the vast majority have a low level of the professional competence of future electrical engineering teachers; the main reason for the low level of the professional competence of future electrical engineering teachers is the lack of mechanisms for their involvement in electrical-based activities in the classroom. The proposed and implemented methodical system of future teachers' training within the formative stage of the pedagogical experiment allowed freedom of choice of courses depending on the social demand for highly qualified pedagogical workers, self-expression of talented students. It ensured the implementation of the methodology of teaching and research - the latest educational models of training of tutors, coaches, facilitators, advisors, mentors, and coaches. Thus, the methodical system of the professional competence of future electrical engineering teachers has been successfully tested at the formative stage. The analysis of the obtained data demonstrated its effectiveness and the need to introduce it in higher education pedagogical institutions.

The findings enabled the following conclusions: the presented methodical system of development of the professional competence of future electrical engineering teachers gives the possibility to reflect more clearly the interaction not only components of the system but also activities of the learning process participants and particularly their cooperation, mutual influence in the process of realization of the educational purposes. The effectiveness of the proposed methodical system depends on the level of professional competence of the teacher and the influence of the administration. The use of all elements of the methodical system will also ensure the development of students' personality-oriented readiness for teaching. In our further activities, we plan to develop methodical support for the implementation of the methodical system for the development of the professional competence of future electrical engineering teachers, as well as the dissemination of its successful implementation in Ukrainian higher education institutions.

References

1. On Education: The Low of Ukraine, vol. 38–39. (2017). https://zakon.rada.gov.ua/laws/show/2145-19. Accessed 15 Sept 2024
2. Larson, L.M., et al.: The academic environment and faculty satisfactioi die role of psychological needs. J. Career Assess. **27**(1), 1–16 (2017). https://doi.org/10.1016/j.ijnurstu.2013.06.009
3. Juszczyk, S., Kim, S.: Distance learning in the polish and korean universities during COVID-19 pandemic. New Educ. Rev. **62**(4), 115–127 (2020). https://doi.org/10.15804/tner.20.62.4.10
4. Kong, L.N., Qin, B., Zhou, Y.Q., Mou, S.Y., Gao, H.M.: The effectiveness of problem-based learning on development of nursing students' critical thinking: a systematic review and meta-analysis. Int. J. Nurs. Stud. **51**(3), 458–469 (2014). https://doi.org/10.1016/j.ijnurstu.2013.06.009
5. Khampirat, B.: The Impact of work-integrated learning and learning strategies on engineering students' learning outcomes in Thailand: a multiple mediation model of learning experiences and psychological factors. IEEE Access **9**, 111390–111406 (2021). https://doi.org/10.1109/ACCESS.2021.3055620
6. Hamadi, M., El-Den, J., Azam, S., Sriratanaviriyakul, N.: Integrating social media as cooperativelearning tool in higher education classrooms: an empirical study. J. King Saud Univ.-Comput. Inf. Sci. (2021). https://doi.org/10.1016/j.jksuci.2020.12.007
7. Kalenskyi, A., Gerliand, T., Kravets, S., Homeniuk, D., Nagayev, V.: Dual educational system of professional training of future skilled workers. In: Design, Simulation, Manufacturing: The Innovation Exchange, pp. 369–378. Springer, Cham (2023). https://doi.org/10.1007/978-3-031-32767-4_35

8. European Training Strategy. Set of competences for trainers. Glossary/SALTO Training and Cooperation Resource Centre. https://www.saltoyouth.net/downloads/4-17-3045/Glossary%20-%20Set_of_trainer_competences_low%20resolution.pdf. Accessed 13 Feb 2024. Accessed 15 Sept 2024
9. Behavioural Competency Dictionary University of Guelph (2010). https://www.uoguelph.ca/cio/sites/uoguelph.ca.cio/files/HR's%20Competency%20Dictionar.pdf. Accessed 13 Feb 2024. Accessed 15 Sept 2024
10. Luzan, P., Titova, O., Kurok, R., Mosia I.: The methodology for assessment of engineering students outcomes. In: 2021 IEEE International Conference on Modern Electrical and Energy Systems (MEES), Kremenchuk, Ukraine, pp. 1–6 (2022). https://doi.org/10.1109/MEES52427.2021.9598666
11. Titova, O., Luzan, P., Yarosh, L., Symonenko S., Zaitseva, N.: The test instruments development based on the items complexity. In: 2022 IEEE 4th International Conference on Modern Electrical and Energy System (MEES), Kremenchuk, Ukraine, pp. 1–6 (2022) doi.org/ https://doi.org/10.1109/MEES58014.2022.10005697
12. Svyrydiuk, V., Luzan, P., Svyrydiuk, O., Titova, O., Popova, O.: Electronic textbooks as means for developing professional competencies of engineering students. In: Tonkonogyi, V., Ivanov, V., Trojanowska, J., Oborskyi, G., Pavlenko, I. (eds.) Advanced Manufacturing Processes III. InterPartner LNME, 2021, pp. 707–716. Springer, Cham (2022). https://doi.org/10.1007/978-3-030-91327-4_68
13. Bykov, V.: Methodological systems of modern information and educational technologies. Problems and prospects of the formation of the national humanitarian and technical elite: Collection of science works, vol. 3, pp. 73–83 (2002). https://lib.iitta.gov.ua/502/1/Bykov4-2002.pdf. Accessed 13 Feb 2023. Accessed 15 Sept 2024
14. Nagayev, V., Gerliand, T., Kyrepin, V., Nagayeva, G., Sosnytska, N., Yablunovska, K.: Pedagogical technology of management of students' educational and creative activities in the process of professional training of engineers. In: International Conference on Modern Electrical and Energy Systems (MEES), pp. 1–4. IEEE (2021)
15. Titova, O., Luzan, P., Sosnytska, N., Kulieshov, S., Suprun, O.: Information and communication technology tools for enhancing engineering students' creativity. In: Advances in Design, Simulation and Manufacturing IV. DSMIE 2021. Lecture Notes in Mechanical Engineering, pp. 332–340. Springer, Cham (2021). https://doi.org/10.1007/978-3-030-77719-7_33
16. Kalenskyi, A., Gerliand, T., Lapa, O., Pyatnichuk, T., Haiduk, O.: Implementation of the methodical system for the development and application of eco-oriented pedagogical technologies. In: Ivanov, V., Trojanowska, J., Pavlenko, I., Rauch, E., Piteľ, J. (eds.) Advances in Design, Simulation and Manufacturing VII. DSMIE 2024. Lecture Notes in Mechanical Engineering, pp. 498–508. Springer, Cham (2024). https://doi.org/10.1007/978-3-031-61797-3_42
17. Belu, R.: Project-based teaching approach of a combined undergraduate and graduate course in power electronics. In: Proceedings of the 2019 ASEE Annual Conference and Exposition, Tampa, Florida (CD) (2019). https://www.researchgate.net/publication/344797299_Board_69_Project-based_Teaching_Approach_of_a_Combined_Undergraduate_and_Graduate_Course_in_Power_Electronics. Accessed 15 Sept 2024
18. Hosseinzadeh, N., Hesamzadeh, M.R.: Application of project-based learning (PBL) to the teaching of electrical power systems engineering. IEEE Trans. Educ. **55**(4), 495–501 (2012). https://doi.org/10.1109/TE.2012.2191588
19. Pavliuk, L.: Features of the formation of a methodological system for the training of future technology teachers. Professional Pedagogy **1**(24), 61–66 (2022). https://doi.org/10.32835/2707-3092.2022.24.119-126

20. Markova, O.M.: The model of the methodical system and the goals of learning the basics of mathematical informatics for students of technical universities. Herald of Cherkasy University. Pedagogical Sciences Series, vol. 7, pp. 36–42 (2016). https://doi.org/10.48550/arXiv.1807.08937
21. Kotsan, I.: Professional training of a modern teacher: problems and guidelines. Higher Education of Ukraine, vol. 2, pp. 13–19 (2013). http://nbuv.gov.ua/UJRN/vou_2013_2_6. Accessed 13 Feb 2024
22. Hadgraft, R., Kolmos, A.: Emerging learning environments in engineering education. Australas. J. Eng. Educ. **25**(1), 3–16 (2020). https://doi.org/10.1080/22054952.2020.1713522

The Development of Engineering Teachers' Environmental Competence

Oleksandr Koshuk[1], Olena Titova[2](✉), Petro Luzan[2], Iryna Mosia[2], and Tetiana Zhukova[3]

[1] National University of Life and Environmental Sciences of Ukraine, 15, Heroiv Oborony St., Kyiv 03041, Ukraine

[2] Institute of Vocational Education of the National Academy of Educational Sciences of Ukraine, 98a, Vito-Lytovskyi Lane, Kyiv 03045, Ukraine
olena.titova@tsatu.edu.ua

[3] Dmytro Motornyi Tavria State Agrotechnological University, 18, B. Khmelnitsky Ave., Melitopol 72312, Ukraine

Abstract. The current global environmental crisis and the significant negative impacts of technology and industry on the environment highlight the urgent need for transformative changes in engineering education. Despite increasing awareness of sustainability issues, existing engineering curricula often lack a structured approach to cultivating graduates' environmental awareness, understanding of ecological principles, and adherence to environmental safety standards. This gap limits engineers' capacity to address critical challenges related to sustainable development. Engineering teachers play a pivotal role in bridging this gap, as they are responsible for equipping students with the skills and knowledge required to integrate sustainability into professional practices. However, there is a notable lack of comprehensive frameworks and tools to enhance the environmental competence of engineering educators. This study addresses this challenge by identifying and substantiating the goals, content, and effective methodologies for fostering environmental competence in engineering teachers. Grounded in advanced principles of environmental education, the research introduces a combined model for integrating sustainability into engineering curricula and an algorithm for greening discipline content. By implementing these innovative approaches, the study provides examples of cases to empower engineering educators to prepare environmentally competent engineers capable of addressing ecological challenges and promoting sustainable production practices.

Keywords: Environmental Education · Engineering Students · Teacher's Professional Competence

1 Introduction

The purposeful development of environmental knowledge, skills, and abilities among individuals and professionals is crucial in addressing the global environmental crisis, which continues to escalate worldwide. While technological progress drives societal

F. J. G. Silva et al. (Eds.): DSMIE 2025, LNME, pp. 148–159, 2025.
https://doi.org/10.1007/978-3-031-96413-8_14

advancement, it also exerts a significant toll on the environment. The technology sector is a growing contributor to global emissions due to the increasing demand for machine building, data centres, cloud computing, and electronic devices. High energy consumption, coupled with the extraction and processing of essential materials, further exacerbates this impact, often leading to land and water degradation. E-waste, comprising discarded electronics, adds to this challenge, with approximately 57.4 million metric tons generated globally in 2023 [1]. As highlighted in [2], the need for environmentally responsible production methods is critical to mitigating these effects. Addressing these issues requires collective action from industries, governments and individuals to integrate sustainable practices across all levels.

While society's evolving mindset recognizes the importance of rejecting numerous past practices to protect the environment, significant negative impacts remain within the technology sector that still need to be addressed. The obvious way out is environmental regulation [3] which requires replenishment of enterprises and organizations with environmentally competent specialists, in particular technical staff and engineers. That includes thorough professional and environmental training, developed ecological thinking and high environmental awareness. On the other hand, the educational process is not yet sufficiently focused on the environmentalization of the engineering education content [4]. This problem is actualized by the fact that usually, the environmental education of technicians, technologists and engineers refers to the basics of environmental protection [6], though the issue has been studied for more than 20 years [6, 7]. Numerous studies [8] show that nowadays individuals with a degree are more inclined to engage in pro-environmental behaviours across various contexts and are more ready to work under conditions of a circular economy and European Green Deal [9].

Those concepts require the improvement of environmental competence and professional training of engineers that must be provided by general mandatory comprehensive education in the field of environmental protection for engineers [9] or as a separate direction - Environmental Engineering [11]. We also support the idea that environmental knowledge is an integral qualification requirement for all officials whose activities are related to the use of natural resources and lead to impact [12]. These provisions directly apply to engineering teachers, as every teacher involved in the educational program must use the tools of his discipline to purposefully prepare students for effective resolution of professional and environmental situations in the future. Although, the practice shows that, unfortunately, engineering teachers are not actively involved in this process [12, 13]. We support the critics of the current environmental situation in society and the proposal to develop environmentally friendly technical systems based on process automation presented in [14]. Additionally, in the European Union, the development of teachers' environmental competence is recognized as essential for promoting sustainability education [15, 16]. That supported the novelty of the current research which was to study the issue widely for the development of some definite concepts and recommendations for engineering teachers. Moreover, the issue has not been studied widely for the development of some definite concepts and recommendations for engineering teachers.

Therefore, to increase the level of engineering students' environmental and professional training, the educational process must be significantly greened (according to [17] greening means that "the person or organization is becoming more aware of environmental issues"), ensuring, in particular, the dynamic development of teachers' environmental competence using appropriate methods.

Thus, the current research aimed to substantiate some tools for the development of engineering teachers' environmental competence based on the advanced ideas, provisions and principles of environmental education and environmentalization of the educational process. To achieve the goal, the following tasks were set:

- To substantiate the goals and content for the development of engineering teachers' environmental competence;
- To design a combined model of environmental education for engineering students;
- To develop an algorithm for environmentalization of the engineering discipline content.

2 Literature Review

The European Union (EU) incorporates various legal, economic, and social foundations for environmental protection through a combination of key normative documents and strategies. Environmental protection in the EU is enshrined in foundational documents such as the Charter of Fundamental Rights of the European Union, which explicitly recognizes the right to a high level of environmental protection under Article 37 [18]. Additionally, EU directives such as the Waste Framework Directive (2008/98/EC) [19] and the Environmental Impact Assessment Directive (2011/92/EU) [20] provide clear guidelines for environmental management, emphasizing the integration of environmental concerns into planning and policy-making across member states. These directives form a legal framework that not only ensures compliance but also promotes environmental awareness among citizens and institutions.

The European Green Deal [9], one of the EU's most ambitious strategies, offers a comprehensive roadmap to achieve carbon neutrality by 2050. It promotes sustainable development through green technologies, circular economy initiatives and financial incentives. Coupled with the EU Emissions Trading System (ETS) [21], these economic tools encourage industries to reduce greenhouse gas emissions while fostering innovation. Furthermore, the EU's solidarity principle [22] and mechanisms like the European Social Fund [23] address the social impacts of the green transition, ensuring inclusivity and fairness, especially for workers in sectors affected by these changes.

While these frameworks establish an extensive foundation, their direct application to engineering education, particularly in developing environmental competence among teachers, remains less explored. Previous research [4, 9, 24] emphasizes that integrating environmental education into engineering curricula can significantly enhance sustainability competence among students. However, these studies also reveal gaps in the systematic training of engineering teachers, who play a pivotal role in fostering ecological mindsets. For instance, Li et al. (2023) [24] demonstrate the effectiveness of community interaction and open-source learning in sustainability competence development, highlighting the need for educators to adopt innovative teaching practices.

The necessity for the purposeful development of environmental competence in engineering teachers is crucial. In the previous research [25, 26] the model for professional competence development, identifying the multi-stage and complex nature of this process was proposed. Moreover, in [26] the importance of equipping general education teachers with environmental competencies to support the training of environmentally conscious engineers was underscored.

The mentioned findings align with the current study's objective to design a systematic approach for the development of engineering teachers' environmental competence.

The content of EU policies and directives, as well as insights from recent studies, directly informs the proposed model and algorithm by emphasizing the importance of embedding environmental concepts into all aspects of engineering education. The "greening" of curricula (saturating educational content with environmental protection material) is a key strategy supported by existing research. This approach ensures the consistent development of students' environmental knowledge, ecological thinking, and ethical perspectives. Furthermore, the role of educators in this process cannot be overstated. Their ability to integrate environmental considerations into technical disciplines is directly tied to the success of such initiatives.

While the EU provides a robust policy framework for environmental protection, its application to engineering teacher training requires further exploration. The reviewed literature highlights significant progress and persistent gaps in this area. The current study builds upon these findings, proposing a structured approach to enhance the environmental competence of engineering educators, thereby addressing both professional and ecological challenges.

3 Research Methodology

In the first stage of the research, to clarify the level of development of the research problem, determine the directions of scientific exploration and identify the features of the development of the engineering teachers' environmental competence, theoretical analysis of scientific works, a study of the practical reports, requirements of normative and legal documents of an environmental nature were provided.

Then the term "environmental competence" was clarified. Therefore, in the current research we considered the environmental competence as a person's ability to apply environmental knowledge and experience in professional and life situations, guided by the priority of environmental values and the non-pragmatic motivation of interaction with the environment, based on the awareness of personal involvement in environment protection problems and responsibility for the environmental consequences of one's professional and everyday activities. That allowed to substantiate the goals and content for the development of engineering teachers' environmental competence highlighting in the structure of the engineering teacher professional competence the environmental component as the ability of an individual in specific environmental situations to actualize the acquired knowledge and experience of a valuable attitude to nature to make environmentally appropriate decisions, to responsibly solve life situations, subordinating the satisfaction of one's needs to the principles of sustainable development of society, promoting one's socially significant environmental position and purposefully carry out environmental training of students.

Thus, environmental competence was understood as: the ability of an individual to make environmentally safe decisions and act without harming the environment; the ability to demonstrate environmental culture in the "zone of responsibility". Moreover, it is a characteristic that enables the teacher to responsibly regulate life situations, subordinating the satisfaction of one's own needs to the principles of sustainable development; the teacher's ability for situational activity in the natural environment.

Then to define the level of the development of engineering teachers' environmental competence the engineering teaching practice was studied. A questionnaire for engineering teachers' self-assessments was provided. The engineering teachers from Ukrainian technical colleges were offered to assess and compare their levels of environmental and educational competencies development. The self-assessment was carried out on a 5-point scale in the following order: if the respondent believed that he had a "Master" level of development of educational, environmental, etc. competences, it was marked with a score of 5; other marks were 4 for "Experienced", 3 for the "Competent Practitioner", 2 for the "Intermediate Beginner (Emerging Practitioner)" and 1 for a "Beginner (Novice)".

With a total population of approximately 5050 engineering teachers at Ukrainian colleges, the sample of 477 respondents was considered representative. The average scores of respondents were as follows: 4.62 and 3.73 according to educational and environmental competences, respectively. A comparison of the distributions of the studied choices using the Wilcoxon-Mann-Whitney test showed that the empirical value of the criterion (12.34) was greater than the critical value (1.96). With a probability of 95%, it could be stated that the obtained data were not random.

Finally, based on the successful practices [3, 5] as well as the idea "to integrate engineering design and the environment" [7], a combined model of environmental education for engineering students and an algorithm for environmentalization of the engineering discipline content were designed as the tools for engineering teachers' environmental competence development in practice.

Using the findings published in [13] the most appropriate forms for environmental education were defined. Interdisciplinary environmental education for engineers emphasizes fostering collaborative teamwork skills, including effective organization and team management. To equip students for real-world environmental challenges, it was essential to implement educational strategies that not only cultivate a robust understanding of multiple disciplines but also highlight the development of cooperative values and problem-solving approaches.

4 Results and Discussion

The current state of the natural environment, coupled with the pressing social and economic challenges, underscores the critical need for a more integrated approach to professional training for engineers. A holistic understanding of environmental concerns, sustainability principles, and resource management is essential for shaping responsible professionals. However, the results of the present study confirmed that one of the key barriers to the effective greening of engineering education lied in the insufficient environmental competence of engineering educators. This finding aligns with the conclusions drawn by several researchers, which highlight the inadequacy of teacher training in environmental issues as a fundamental obstacle to achieving environmental goals in education [10, 27].

The self-assessment results of engineering teachers presented in this study further corroborate this point, as shown in Fig. 1.

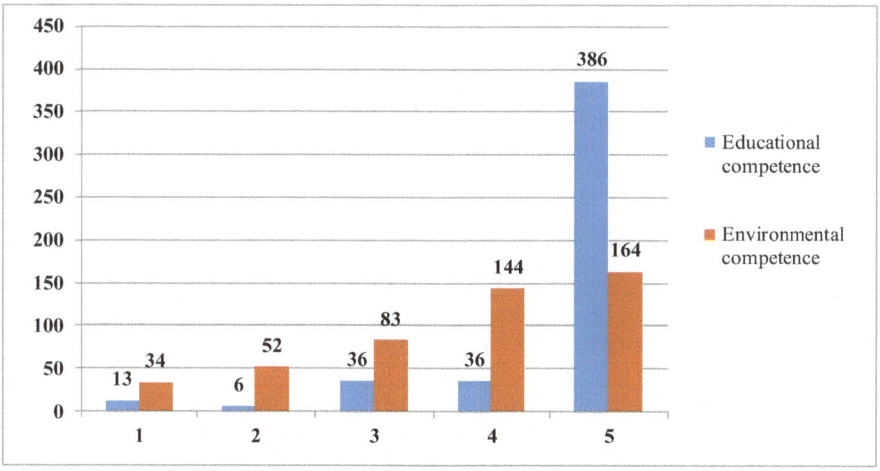

Fig. 1. The results of the engineering teachers' survey regarding the development of their educational and environmental competencies.

While a significant proportion of the teachers (80.9%) rated their educational competence as high, only 34.4% of respondents felt confident in their environmental competence. In contrast, a notable 10.9% reported low environmental competence. These results clearly indicate that, despite possessing strong educational skills, engineering teachers often lack the necessary environmental expertise and the ability to integrate sustainability concepts into their teaching. This disparity is also evident in recent studies, which suggest that engineering faculty are often underprepared to deliver effective environmental education, both in terms of content knowledge and pedagogical strategies [2, 11].

The need for enhancing the environmental competence of engineering educators has been recognized in numerous studies. A study by [3] emphasizes the role of teachers as mediators of knowledge that can drive the adoption of green technologies and sustainability practices in professional fields. Without properly trained educators, students may fail to acquire the necessary skills to meet the challenges of the green economy. In a similar vein, the European Commission's GreenComp framework [15] calls for the development of environmental competences among educators as a prerequisite for embedding sustainability in education.

Furthermore, the combination of environmental and educational competencies within engineering curricula has been proven to be an effective strategy for fostering sustainability-oriented mindsets among students as coincided with [13]. The model developed in this study (Fig. 2) presents a strategic framework for enhancing engineering teachers' environmental competence.

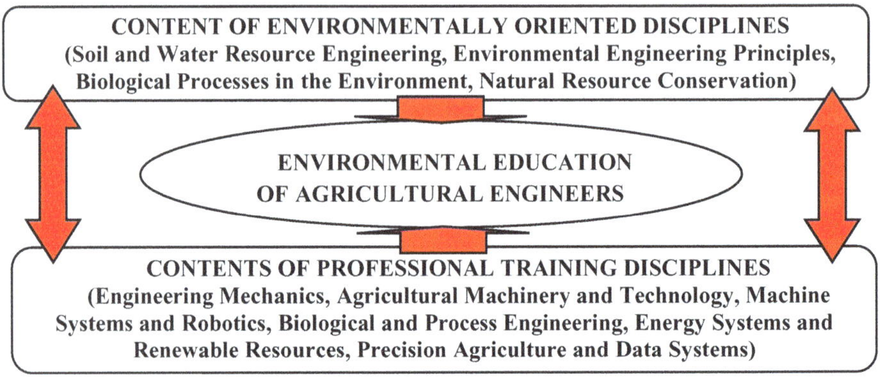

Fig. 2. The combined model of environmental education of agricultural engineers.

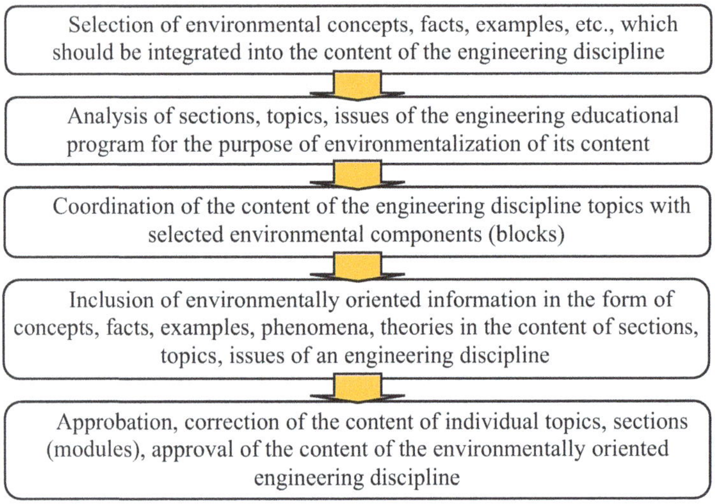

Fig. 3. Sequence of stages for greening the content of a professional discipline.

The analysis of case studies from institutions such as Iowa State University and Purdue University demonstrates that integrating sustainability into engineering curricula not only improves students' understanding of environmental challenges but also leads to the development of key competences that are vital for addressing global sustainability issues. These findings align with those of Bocconi et al. (2016) [27], who argue that the integration of sustainability into curriculum design fosters the development of competencies needed for responsible decision-making in complex environmental contexts. In parallel, the algorithm for the environmentalization of engineering disciplines proposed in this study (Fig. 3) has been validated by successful case studies, as demonstrated by recent research. The methodical integration of environmental principles into engineering education, as outlined in the five-step process, enhances both the depth of environmental knowledge and the teachers' ability to guide students in applying eco-friendly solutions to real-world challenges. Leifler and Dahlin (2020) [10] argue that the incorporation of sustainability into engineering curricula must be systematic, involving the careful selection and integration of relevant environmental concepts, as well as continuous feedback from students and colleagues to ensure the effectiveness of the educational approach. This is particularly important in a rapidly evolving field where new environmental challenges and technological innovations are constantly emerging.

The key advantage of the proposed algorithm was its emphasis on gradual integration, which allows teachers to expand their environmental expertise step by step. This progressive approach, coupled with active feedback mechanisms, is critical for overcoming barriers to the environmentalization of engineering education. This approach was similar to that recommended by Huaquisto-Cáceres et al. (2022) [5], who advocated for a phased integration of environmental education within engineering curricula to foster deeper engagement with sustainability concepts.

The role of teachers in the environmental education of engineers extends beyond merely transferring knowledge; they must also model sustainable practices and integrate interdisciplinary perspectives. This aligns with the findings of Li et al. (2023) [24], who stress that fostering individual sustainability competence requires community interaction and open-source learning. In this context, the teacher's environmental competence evolves not only through their own learning but also through their engagement with students, colleagues, and the broader community of sustainability advocates.

Moreover, the algorithm for greening the content of engineering curricula not only enhances the environmental knowledge of students but also supports the development of teachers' professional competences. By actively engaging in curriculum analysis, teachers sharpen their skills in interdisciplinary thinking, aligning engineering principles with environmental challenges.

Here are some examples of the environmentally oriented content for the agricultural engineering curriculum disciplines "Soil and Water Resource Engineering" (1) and "Agricultural Machinery and Technology" (2).

1. Topic "Sustainable Soil and Water Management Strategies".

Understanding the environmental impacts of soil and water management is essential in engineering practices. This topic emphasizes the importance of reducing soil erosion through conservation tillage, implementing contour farming, and establishing riparian buffer zones to prevent nutrient runoff. Engineers must design water management systems, such as rainwater harvesting and precision irrigation to optimize water use while minimizing waste and depletion of natural aquifers.

Case studies highlight the integration of renewable energy in water pumping systems and the use of bioengineered solutions, like vegetative swales to manage stormwater sustainably. Special attention is given to the assessment of soil health using organic matter and microbial activity as indicators promoting practices that enhance carbon sequestration and biodiversity.

This environmentally oriented approach ensures that engineering solutions contribute to long-term ecosystem stability and resilience against climate change.

2. Topic "Sustainable Mechanization and Green Technologies in Agriculture"

This topic focuses on environmentally friendly innovations in agricultural machinery to promote sustainable farming practices. Key concepts include the development and use of low-emission tractors and implements powered by renewable energy sources like solar or biofuels.

Students analyse precision agriculture technologies, such as GPS-guided machinery, which optimize planting, fertilizing, and harvesting processes to reduce resource wastage and environmental impact. Discussions cover soil-friendly machinery like reduced-tillage equipment designed to preserve soil structure and organic matter.

Case studies explore advancements in automation and robotics for efficient resource utilization and minimal ecological disturbance. Emphasis is placed on lifecycle analysis of machinery to assess sustainability from manufacturing to disposal.

Practical sessions may include designing or retrofitting agricultural machinery to meet eco-efficiency goals aligning engineering practices with global sustainability standards in agriculture.

This integration of sustainability into teaching practices aligns with the recommendations of the European Commission's GreenComp framework [15], which stresses the need for educators to possess the skills necessary to guide students in understanding and addressing the complex environmental challenges they will face in their professional careers.

These examples are consistent with the findings of Van den Beemt et al. (2020) [13], who emphasize the importance of embedding sustainability concepts within the technical disciplines to ensure that engineers are prepared to tackle the challenges of a green economy.

The results of this study also suggested that the inclusion of interactive learning technologies, such as cooperative learning, gamification, and problem-solving scenarios, was essential for promoting deeper engagement with environmental topics. These technologies have been shown to enhance student motivation and foster a collaborative learning environment that is conducive to developing sustainability competencies [27]. According to [28], fostering such an environment is crucial for the development of students' critical thinking and problem-solving skills, which are necessary for addressing real-world environmental challenges.

Finally, the environmentalization of engineering curricula, while crucial, is not sufficient on its own to ensure that students acquire the necessary competencies for sustainable development. As noted by Woolard et al. (2022) [11], the integration of sustainability into engineering education must be complemented by the development of students' ecological thinking, awareness, and practical skills. This requires a shift towards more interactive and student-centered teaching methods, which actively involve students in the learning process and encourage them to apply environmental principles to real-world scenarios. These findings support the conclusions of Li et al. (2023) [24], who argue that the development of sustainability competence in engineering education should be viewed as an ongoing process, with continuous feedback and adaptation to new challenges and opportunities.

The integration of environmental education into engineering curricula is a multifaceted process that requires the active involvement of teachers, students, and educational institutions. The results of this study confirmed that engineering educators' environmental competence must be enhanced through targeted training and the systematic integration of sustainability principles into the curriculum. The proposed algorithm for environmentalizing engineering disciplines could provide a structured framework for achieving this goal, and the successful case studies presented in this study further highlight its potential for improving both teaching practices and student learning outcomes. Ultimately, the success of environmental education in engineering depends on the continued collaboration between educators, researchers, and policymakers to create a more sustainable future.

Environmental education of college teachers should be carried out continuously, including in the inter-course period of advanced training. Environmental knowledge is constantly being updated, new environmental areas are emerging: environmental management, psychoecology, ethnic ecology, urban ecology, radioecology, etc. Therefore, each college should have an effective methodological system that organizes and ensures continuous environmental education of teachers by means of advanced training courses, forms of corporate training.

5 Conclusions

This research focused on the critical role of engineering teachers' environmental competence in the effectiveness of environmental education for engineering students. The study emphasized that teachers' ability to integrate sustainability into their teaching is essential for fostering environmental awareness and sustainable decision-making in future engineers. The findings from the survey of 477 Ukrainian engineering teachers revealed that only 34.4% demonstrated a high level of environmental competence, highlighting a significant gap in the current education system. Therefore, all teachers of professional disciplines should be involved in the environmental education of future technicians, technologists and engineers. To this end, the content of professional training courses should be eco-friendly, and students should be offered eco-oriented learning technologies.

To address this challenge, a combined model for environmental education and an algorithm for greening engineering curricula were developed. These tools aim to enhance the environmental competence of engineering teachers, enabling them to incorporate sustainability principles into their courses effectively. The proposed model and algorithm provide a systematic, interdisciplinary approach to integrating environmental knowledge into engineering disciplines, facilitating teachers' continuous development.

Future research should focus on creating a comprehensive methodological system that combines formal, non-formal, and informal education strategies. This system would support the ongoing development of environmental knowledge, ecological thinking, and sustainability practices for both engineering teachers and students, ensuring the advancement of technical education aligned with sustainability goals.

References

1. GreenMatch: Technology and Its Environmental Impact. GreenMatch Blog. https://www.greenmatch.co.uk/blog/technology-environmental-impact. Accessed 26 Nov 2024
2. Silva, F.J.G., Gouveis, R.M.: Cleaner Production - Toward a Better Future. Springer (2020)
3. Du, K., Cheng, Y., Yao, X.: Environmental regulation, green technology innovation, and industrial structure upgrading: the road to the green transformation of Chinese cities. Energy Econ. **98**, 105247 (2021). https://doi.org/10.1016/j.eneco.2021.105247
4. Borges, M.G.: Environmentalization. In: Gooren, H. (ed.) Encyclopedia of Latin American Religions, pp. 1–5. Springer, Cham (2015). https://doi.org/10.1007/978-3-319-08956-0_24-1
5. Huaquisto-Cáceres, S., Hurtado-Chávez, E., Chambilla-Flores, I.G., Barrientos-Paredes, K.N.: Environmental education and ethics in civil engineering study programs in the Puno region, Peru. DYNA **89**(222), 18–27 (2022). https://doi.org/10.15446/dyna.v89n222.101210
6. Abdul-Wahab, S.A., Abdulraheem, M.Y., Hutchinson, M.: The need for inclusion of environmental education in undergraduate engineering curricula. Int. J. Sustain. High. Educ. **4**(2), 126–137 (2003). https://doi.org/10.1108/14676370310467140
7. Jahan, K., Everett, J., Hesketh, R., Jansson, P., Hollar, K.: Environmental education for all engineers. Water Sci. Technol. **49**, 19–25 (2004). https://doi.org/10.2166/wst.2004.0478
8. Ramu, S., Priyanka, V., Srilatha, K.: Role of environmental education in engineering curriculum. Int. J. Sci. Prog. Res. (IJSPR) **41**(2), 69–73 (2017). https://www.ijspr.com
9. European Commission: The European Green Deal. https://commission.europa.eu/strategy-and-policy/priorities-2019-2024/european-green-deal_en. Accessed 27 Nov 2024
10. Leifler, O., Dahlin, J.-E.: Curriculum integration of sustainability in engineering education – a national study of programme director perspectives. Int. J. Sustain. High. Educ. **21**(5), 877–894 (2020). https://doi.org/10.1108/IJSHE-09-2019-0286
11. Woolard, C.R., et al.: Developing an integrated environmental engineering curriculum. Am. Soc. Eng. Educ. (2022)
12. Obrecht, M., Feodorova, Z., Rosi, M.: Assessment of environmental sustainability integration into higher education for future experts and leaders. J. Environ. Manag. **316**, 115223 (2022). https://doi.org/10.1016/j.jenvman.2022.115223
13. Silva, F.J.G.: Green Automation: Increasing Sustainability, From Industry to Our Home (2024)
14. European Commission, Joint Research Centre: GreenComp: A New Competence Framework for Learning and Teaching Sustainability. https://joint-research-centre.ec.europa.eu/jrc-news-and-updates/jrc-publishes-new-competence-framework-support-learning-environmental-sustainability-2022-01-14_en (2022)
15. Uni-T Academy: TEAcher COmpetence for Sustainability and Environmental Education (TEACOSEE). In: Uni-T Academy (ed.) TEACOSEE Project (2022). https://uni-t-academy.eu/research-action-projects/teacosee
16. Collins Dictionary: Greening (2024). https://www.collinsdictionary.com/dictionary/english/greening
17. Van den Beemt, A., et al.: Interdisciplinary engineering education: a review of vision, teaching, and support. J. Eng. Educ. **109**(3), 508–555 (2020). https://doi.org/10.1002/jee.20347
18. Union European Union: Charter of Fundamental Rights of the European Union. Official Journal of the European Union, C 326 (2012). https://www.europarl.europa.eu/charter/pdf/text_en.pdf. Accessed 27 Nov 2024
19. European Parliament and Council of the European Union: Waste Framework Directive 2008/98/EC. Official Journal of the European Union, L 312 (2008). https://eur-lex.europa.eu/legal-content/EN/TXT/?uri=CELEX%3A32008L0098. Accessed 27 Nov 2024
20. European Parliament and Council of the European Union: Environmental Impact Assessment Directive 2011/92/EU. Official Journal of the European Union, L 26 (2012). https://eur-lex.europa.eu/legal-content/EN/TXT/?uri=CELEX%3A32011L0092. Accessed 27 Nov 2024

21. European Parliament and Council of the European Union: Directive 2003/87/EC establishing a scheme for greenhouse gas emission allowance trading within the European Union. Official Journal of the European Union, L 275 (2003). https://eur-lex.europa.eu/legal-content/EN/TXT/?uri=CELEX%3A32003L0087. Accessed 27 Nov 2024
22. Eurofound: Solidarity Principle. European Industrial Relations Dictionary. Dublin (2011). https://www.eurofound.europa.eu/en/european-industrial-relations-dictionary/solidarity-principle. Accessed 27 Nov 2024
23. European Commission: European Social Fund (ESF). https://ec.europa.eu/esf/home.jsp. Accessed 27 Nov 2024
24. Li, H., et al.: Individual sustainability competence development in engineering education: community interaction open-source learning. PLoS ONE **18**(11), e0294421 (2023). https://doi.org/10.1371/journal.pone.0294421
25. Luzan, P., Titova, O., Mosia, I., Pashchenko, T., Navruzova, G.: The model for professional competence development of engineering teachers at colleges. In: Ivanov, V., Trojanowska, J., Pavlenko, I., Rauch, E., Piteľ, J. (eds.) Advances in Design, Simulation and Manufacturing VII. DSMIE 2024. Lecture Notes in Mechanical Engineering, pp. 509–520. Springer, Cham (2024). https://doi.org/10.1007/978-3-031-61797-3_43
26. Titova, O., Luzan, P., Ishchenko, T., Kabysh, M., Homeniuk, D.: Improvement of professional competence of general education teachers for engineering curriculum. In: Tonkonogyi, V., Ivanov, V., Trojanowska, J., Oborskyi, G., Pavlenko, I. (eds.) Advanced Manufacturing Processes V. InterPartner 2023. Lecture Notes in Mechanical Engineering. Springer, Cham (2024). https://doi.org/10.1007/978-3-031-42778-7_35
27. Bocconi, S., Brunello, A., Sessa, V.: Learning and teaching for the 21st century: the role of interactive and cooperative learning technologies. In: De Vries, M., Schutte, J. (eds.) Advances in Educational Technology, pp. 101–115. Springer, Cham (2016). https://doi.org/10.1007/978-3-319-46487-4_7
28. Eurydice: Learning for Sustainability in Europe: Building Competences and Supporting Teachers and Schools (2023). https://www.dzs.cz/sites/default/files/2024-04/Learning_for_sustainability_in_Europe_Building_competences_and_supporting_teachers_and_schools.pdf

Formation of the Economic Culture of Transportation Technology Engineers Using Communication Simulations

Viktor Nagayev(✉), Yuliia Sahachko, Galyna Nagayeva, Svitlana Kuskova, and Yevhenii Beznos

State Biotechnological University, 44, Alchevskih St., Kharkiv 61002, Ukraine
nagaevviktor1966@gmail.com

Abstract. The article aims to substantiate the didactic means of forming the economic culture of transport technology engineers based on the technologies of simulation modeling of production processes. The study used modern methodological approaches, theoretical and empirical methods: questionnaires; testing; simulation modeling; pedagogical experiment, statistical analysis (to determine the level of formation of the engineer's economic culture in the structure of his professional competence), which allow the formation of axiological and economic, and cultural knowledge. The theoretical and methodological aspects of forming the economic culture of managers, because of the methodology of simulation modeling of production situations in the process of their professional training, are considered. The components of the economic culture of a modern transport technology engineer, which are determined by motivational, axiological, activity, and developmental aspects, are substantiated. A method of simulation modeling of production situations is proposed, which helps to form an autodidactic educational environment, optimize organizational forms of practice-oriented personnel training, and ultimately ensures modeling of the economic culture of an engineer. Didactic factors of management of the process of formation of the economic culture of future engineers are determined. Technological means of communication simulations are substantiated based on integration into the content of professional training of educational disciplines of economic and cultural knowledge.

Keywords: economic culture of an engineer · communication simulations · simulation modeling methodology

1 Introduction

The dynamic development of modern economic thinking in the transport logistics system is determined by the humanistic paradigm, which reflects the cultural orientation of engineering personnel training. This is why forming the economic culture of transport technology specialists is of fundamental importance. This pedagogical, economic, and social problem is particularly acute at the level of professional training of engineers in transport technologies in the educational environment, where an insufficient level of

F. J. G. Silva et al. (Eds.): DSMIE 2025, LNME, pp. 160–169, 2025.
https://doi.org/10.1007/978-3-031-96413-8_15

axiological orientation of this process is noted. Future transport technology engineers must be armed with mechanisms of economic orientation of logistics processes, methods of marketing strategies, and methods of optimizing technological costs in production activities, which determines the relevance of this pedagogical problem and the need to form their economic culture. These provisions prove the need to search for innovative pedagogical technologies, forms, and methods of forming the economic culture of future transport technology engineers as an educational result in the context of the axiological dominance of digital pedagogy [1].

The theory and methodology of professional education have accumulated sufficient experience in developing the cultural skills of engineering specialists. However, the modernization issues of didactic renewal of organizational forms, methods, and means of this sphere require the introduction of communication mechanisms, considering modern education's digital transformations. One of the possible approaches to implementing a didactic system for forming an economic culture of future engineers in transport technologies is the use of communication simulations based on simulation modeling. At the same time, as the practice of engineering education proves, transport technology specialists receive methodological tools for developing cultural skills in realistically created simulation models of management situations [2].

Such an approach allows for increasing the productivity of professional training of future specialists and the qualitative level of components of economic culture, since the educational process takes place in an interactive mode of direct communication with subjects of the production sphere. In addition, modern production situations are implemented in the initial process, which are modeled on the plane of real practice and outlined by digital means of communication [3]. Of course, these aspects determine the relevance of the problem under study. They also outline several problematic aspects concerning modeling the educational process in remote access, technological support for achieving educational results, the introduction of reliable technical means of communication simulations, etc. Analysis and methodological substantiation of these pedagogical aspects allows us to outline the scientific scope of this research topic.

2 Literature Review

Modern civilization is simultaneously interpreted by many factors of its innovative development. In this system, an important place belongs to transport technologies, which must be provided with technical, technological, economic, and social components. The importance of the socio-cultural component of logistics operations and transport communications is beyond doubt, since social relations directly or indirectly connect them. This necessitates focusing on economic culture and improving the didactic means of its formation and development in future transport technology engineers [1, 4].

Economic culture, in a broad sense, is a component of general culture. At the same time, it acts as a specific form of the mediation of economic values in a particular sphere of activity. The interpretation of the concept of "culture" is distinguished by its systematic descriptiveness, which testifies to the multifacetedness and unstructuredness of its individual subsystems. Glushman T. examines the concept of "culture of professional activity" as a high level of professionalism of an individual, a way of his creative self-realization, the process of mastering professional values, the formation of a system of moral characteristics, which are considered as the degree of quality of professional activity [5]. Economic culture is a type of general professional culture of a specialist, which corresponds to the block of axiological aspects of the socio-economic responsibility of an individual and an organization in a particular sphere of activity [6].

Economic culture is a multidimensional phenomenon consisting of interacting rational and irrational components. It contains fundamental economic values and norms formed in a given society (professional sphere), and economic laws, ideas, concepts, traditions, certain attitudes and orientations of people regarding the existing economic system and the principles of the relationship between the individual and economic institutions. Prystupa V. The replacement of these components with socio-economic, national-cultural, historical-historical, and mental factors is characterized by stability and is not susceptible to rapid transformational changes in married life [7].

The results of the analysis of the semantic content of this category in most studies mediate economic culture as a specific projection of general culture onto the sphere of economy [8, 9]. Soroka O., Kalaur S. argue that the cultural warehouse of economic activity lies directly as a sign of the professional sphere of activity, and its thought process, which means it is possible to actively integrate into it, activate economic development [8]. The standard of higher education in the specialty 275 "Transport Technologies" provides for the following professional qualities of future specialists as general competencies [1]:

- the ability to preserve and multiply the moral, cultural, scientific values and achievements of society based on an understanding of the history and patterns of development of the subject area, its place in the general system of knowledge about nature and society and in the development of society, technology and technologies, to use various types and forms of physical activity for active recreation and leading a healthy lifestyle;
- the ability to preserve and multiply the moral, cultural, scientific values and achievements of society based on an understanding of the history and patterns of development of the subject area, its place in the general system of knowledge about nature and society, and the development of society, technology, and technologies;
- the ability to evaluate the operational, technical-economic, technological, legal, social, and environmental components of the organization of transportation;
- to critically evaluate society's scientific values and achievements in developing transport technologies.

The content of these educational components determines the importance of the cultural component of professional training of specialists and the need to form economic culture as an integral attribute of the professional competence of future transport technology engineers [9]. The formation of the economic culture of future engineers in transport technologies is a component of a broader problem of economic and managerial education - a system of purposeful influence of society on a person to form in him a humanistic worldview, developed economic thinking, and a high culture of economic behavior. In addition to studying the experience of exact sciences, engineering education requires mastering the system of socio-economic knowledge and developing independent economic thinking on this basis. The theoretical foundation for the formation of an engineer's system of economic knowledge is philosophy and political economy. Brovko K, Ternopilska V. in their research emphasize that the formation of economic and engineering thinking at the current stage of digital transformations is based on the ability to think in terms of "efficiency", compare costs and results, and optimize available reserves [10]. It is important to note that economic and engineering knowledge is most fully implemented in practice only when engineering work is based not only on knowledge, skills and abilities, but also on the experience of creative activity, which is the result of a combination of intellectual and heuristic skills and a socially oriented, economic scientific and practical worldview [11].

Economic and engineering education combines socio-economic knowledge, motives for activity, value orientations, and norms of professional activity of an engineer. Mykhailiuk N. emphasizes that the economic culture of a specialist in the context of a managerial approach is a process of purposeful formation of value orientations of employees of the relevant industry, which are designed to scientifically substantiate the resolution of economic contradictions in the professional environment of the system [12]. Under the above conditions, the teacher of economic disciplines must instill in future engineers an interest in economic processes in the engineering field, create a creative educational environment based on the intensification of independent and individual work of applicants in the conditions of assessing production situations based on the principles of communication simulations.

3 Research Methodology

The research methodology is based on specific methodological approaches (systemic, competency-based, humanistic, axiological, cybernetic), methods of practice-oriented training of specialists, and includes the ascertaining, formative, and control stages of the pedagogical experiment.

At the ascertaining stage of the experiment, the level of cultural training of engineers was clarified, an axiological assessment of their motivation for the formation of economic culture was made by studying the value-oriented content of the organization of transport; the level of knowledge, skills and abilities of the humanistic component of the self-development process was determined; the level of formation of qualities inherent in a future transport technology engineer was investigated. Students of the State Biotechnological University of specialty 275 "Transport Technologies" participated in the survey. The role of experts was played by teachers, production workers, graduate students, and postgraduate students, who, during the semester, observed students, conducted questionnaires, surveys, and analyzed the results of self-assessment of the formation of their economic culture.

The formative stage of the experiment aimed to develop a research concept based on the definition of a hypothesis, scientific laws, and principles of forming the economic culture of engineers using communication simulations of a practically oriented environment. Based on the clarification of theoretical and methodological principles, a set of cases, organizational forms, didactic means of simulation modeling, production situations, and the establishment of the quantitative and qualitative composition of the participants of the experiment were substantiated. The following research methods were used at this stage: forecasting, design, and simulation modeling.

At the control stage of the pedagogical experiment, a comparative assessment of the effectiveness of the proposed model of communication simulations was made using the developed methodology for simulation modeling of decision-making in the transport technology system; the reliability of the scientific hypothesis was assessed. At this stage, the following research methods were used: correlation-regression; questionnaire; comparison.

4 Results and Discussion

Taking into account the preliminary analysis of the functional activity of specialists in specialty 275 "Transport Technologies" [1], the following basic requirements for their personal cultural qualities are proposed: communicativeness; ability for creative activity; adequate self-esteem and a high level of reflection and self-development; orientation towards economic activity and tolerant interaction; economic responsibility; economic motivation and motivation to achieve success; organizational Darwinism, the ability to adapt in crisis situations; high level of self-discipline, economic self-control. Given the above [11, 12], it is possible to specify the characteristics that represent the economic culture of engineers in transport technologies:

- the ability to determine the goal and objectives of economic professional communication in the engineering industry;
- the ability to analyze the subject of economic assessment of technical projects, and organize its discussion;
- skills in managing economic relations with clients and business partners;
- possession of etiquette tools to achieve the goal in the transport communications system;
- ability to conduct business conversations, discussions, negotiations to resolve engineering issues in the system of economic relations;
- ability to convince the interlocutor, prove one's point of view, justify the economic efficiency of the proposed logistics projects, and assess economic risks.

The conducted study of the content of the professional activity of specialists in specialty 275 "Transport Technologies" makes it possible to identify the following elements of their economic culture: organizational culture; culture of business communication; corporate culture; speech etiquette of communication; professional ethics; possession of the basic forms of economic planning and analysis; possession of various communication techniques (concluding economic contracts, conducting business conversations, holding problem meetings, negotiations, etc.); ability to create an atmosphere of comfortable business communication, trust and mutual understanding when resolving economic conflicts [13].

Based on the consideration of the structure of the semantic content of the economic culture of transport technology engineers, we have proposed the following stages of the formation of this professional phenomenon:

1. value-motivational (aimed at the awareness and mastery by applicants of the system of economic values, norms, and rules of professional ethics, and on this basis, the development of interest, needs, interests, and motivation for professional activity as a pedagogical condition for the successful implementation of engineering activities);

2. cognitive-applied (reflects the process of mastering professional, axiological knowledge, communication skills and skills of professional interaction, familiarization with the features of the engineer's economic activity, the content of the phenomenon "economic culture" and its elements: organizational culture, communication culture, ethics of business communication, speech etiquette, ethical-business skills and skills, gaining experience in their application in economic relations with various subjects of the production environment, which contributes to their effective assimilation and further application in engineering activities;
3. professional-activity (direct effective mastery of the components of economic culture based on the application of acquired knowledge (professional, axiological, speech-communicative), skills (professional, communicative, skills of professional interaction, ethical-business), cultural qualities (moral, emotional-volitional, intellectual) in conditions of professional interaction of a specialist-engineer in transport technology.

As a result of the study of the specifics of the professional activities of transport technology engineers, criteria for forming the economic culture of future specialists were determined: motivational, cognitive, and operational.

The motivational criterion is associated with the professional identity of the future specialist and characterizes the motivation for professional activity, which involves the presence of economic values, attitudes, interest, motives for professional activity, motivation for professional communication, which is aimed at ensuring effective interaction in the system of economic relations in the field of transport operations.

The cognitive criterion characterizes the systematicity of professional knowledge in the field of economic culture (the essence and functions of the culture of economic activity; structural components of the ethics of business communication; the mechanism of economic operations, etc.); axiological knowledge (about the values, rules and norms of professional ethics, the culture of economic communication in professional activity); communicative knowledge (about the essence, types, functions, means, methods and techniques of professional economic communication and interaction, methods of psychological influence).

The operational criterion reflects the formation of professional skills and abilities in the system of economic culture and professional activity of an engineer in solving economic issues (the ability to analyze, plan, organize and control transport operations; the ability to form effective interaction between subjects of transport communications; skills of economic analysis and generalization of transport operations; the ability to establish "feedback" with partners in a business communication situation; the ability to conduct business polemics and negotiations, adhere to communication etiquette, possess a culture of business speech; the ability to apply various styles and methods of managing conflict situations to solve economic problems in the system of engineering support of transport technologies.

The analysis of the content of educational and professional programs in the specialty 275 "Transport technologies (in road transport)" [1, 14] in the educational environment using the example of the State Biotechnological University of the first and second levels of higher education allowed us to determine the directions of further effective formation of the economic culture of engineering specialists. In this process, it is essential to implement the pedagogical condition - mastery of axiological knowledge by future engineers by directing the content of the general and professional training cycle disciplines to engage in economic values, ethical norms, and rules in professional activity [15]. The organization of the acquisition of the specified knowledge takes place within the framework of a creative educational environment through the mastery of academic disciplines: 1) general training ("Philosophy", "Corporate Culture", "Ukrainian Language for Professional Purposes"); 2) professional training ("Operations Research and Design of Transport Systems", "Organization and Management of Passenger Transportation", "Economics of Transport Transportation", "Accounting and Analytical Practice"). The content of the academic disciplines of professional training of specialists in the specified specialty integrated topics and educational questions that allow the formation of axiological and economic and cultural knowledge: "Ethics of professional training of an engineer", "Analysis of economic activity in transport", "Culture of economic communication of an engineer in transport technologies" (Fig. 1).

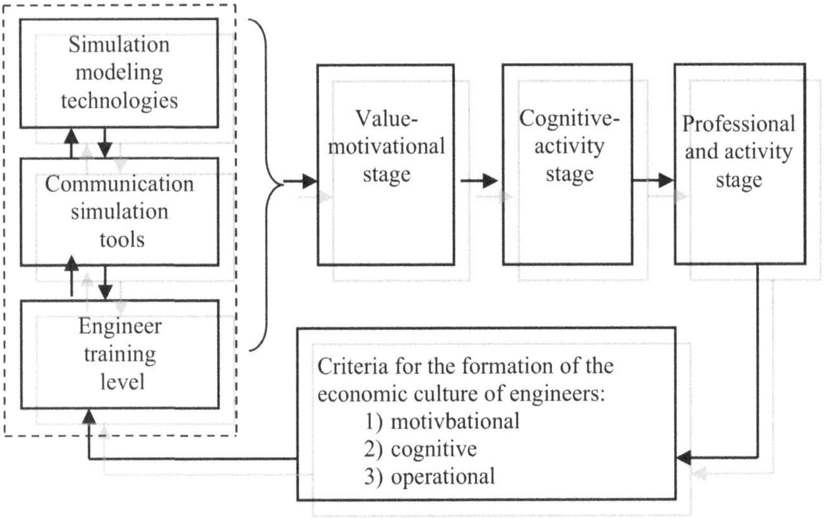

Fig. 1. Methodology for forming the economic culture of transport technology engineers using communication simulations.

In the educational environment of the State Biotechnological University, the process of training specialists in specialty 275 "Transport Technologies" in the formation of economic culture is carried out using a complex of didactic methods (problem, interactive, training, case method, modeling, situational, game) and organizational forms of training (lectures, seminar and practical classes, webinars, trainings, special courses, electives, educational and pre-graduate practice, etc.) [16]. In addition to educational forms, educational aspects of pedagogical influence were activated (student participation in scientific circles, scientific paper competitions, thematic weeks, scientific conferences, professional tournaments for practicing economic skills). For example, in the augmented reality classroom, students can use the Oculus Quest 2 standalone virtual reality headset as a means of mastering spatial virtual models, designing complex 3D models in the Gravity Sketch application, followed by preparing a technical project on a computer in the Ultimaker Cura application and printing it on a Creality Ender 3 S1 Pro 3D printer. This allows you to visualize the result of an engineering solution, develop the spatial boundaries of the professional training content, and improve the humanistic orientation and cultural aspects of the engineer's competency structure.

To help students master the skills of cultural interaction, training technologies and a complex of interactive and dialogic communication teaching methods were used. The development of skills in implementing economic values, norms, and rules of ethics of an engineer was tested based on simulation modeling of communication simulations [17]:

- development and implementation of professionally oriented training "Economic culture of a specialist engineer in transport technologies";
- use of training techniques (information and communication, simulation, coaching, functional and professional) to practice economic culture skills;
- use of a complex of interactive and group forms of learning (situational and business games, work in small groups, case method, communicative situational tasks);
- Use automated complex professional training "Virtual economy of transport operations" using tasks and exercises to develop ethical and business orientation.

There is no discussion about the nutritional optimization of the cultural warehouse at the gas station for the practice-oriented preparation of specialists. Our results demonstrate the effectiveness of the communication simulation technique and, simultaneously, indicate the need for further investigation of the correlational influx of factor signs on the resulting indicators of the formation of an economic culture in the system of professional competence of transport technology engineers.

5 Conclusions

By forming the basic economic culture of future engineers with specialty 275 "Transport Technologies", it is essential to ensure the initial process of preparing these specialists using interactive methods and practical orientation technologies of communication simulations achieves illuminating results. The research results demonstrated the effectiveness of a combination of training technologies and complex interactive and dialogic methods for implementing valuable guidelines for ethical norms of economic culture. The development of training technologies based on the characteristics of simulation modeling of virtual situations (simulation techniques, case technologies, and training of virtual operations) has ensured the intensification of the initial process, increasing the motivation for educational preparation, and forming a cultural background. These pedagogical minds also support the effective professional adaptation of future engineers with transport technologies in the minds of manufacturing.

Educational and methodological support for the process of forming the economic culture of future transport technology engineers has been developed, which is represented by a set of didactic exercises of communication simulations and interactive teaching methods (business game, game and socio-psychological training, simulation modeling method), cultural training based on dialogic and case technologies. The use of simulation tools for communication simulations in the educational process allowed to expand the applicants' understanding of the economic culture of transport technology specialists, to develop their professionally essential qualities of economic culture, business ethical communication skills, skills in substantiating economic procedures of transport technologies, as well as to develop organizational abilities and internal motivation for professional improvement. The proposed didactic approach to implementing communication simulations contributed to forming economic culture qualities in future transport technology engineers as a significant component of general professional competence.

References

1. Higher education standard: Bachelor's degree, field of knowledge 27 Transport, specialty 275 Transport Technologies. Approved by the Order of the Ministry of Education and Science of Ukraine, dated 02/27/2023. No. 213 (2023). https://mon.gov.ua/storage/app/media/Fakhova%20peredvyshcha%20osvita/Zatverdzheni.standarty/2023/02/27/275-Transportni.tekhnolohiyi-za.vydamy-213-27.02.2023.pdf. (in Ukrainian)
2. Nagayev, V., Vertegel, V., Nagayeva, G., Kovalenko, O., Smihunova, O.: A model of formation of professional and creative competences for engineers based on information and digital technologies. In: Ivanov, V., Trojanowska, J., Pavlenko, I., Rauch, E., Piteľ, J. (eds.) Advances in Design, Simulation and Manufacturing VI. DSMIE 2023. Lecture Notes in Mechanical Engineering, pp. 390–399 Springer, Cham (2023). https://doi.org/10.1007/978-3-031-32767-4_37
3. Mykhailiuk, N.V.: The influence of the educational environment on the formation of the professional culture of future bachelors of banking. Modern aspects of science. Svazek XXI International Collective Monograph, vol. 29, pp. 291–328 (2023). (in Ukrainian)
4. Iaresko, K.V.: Culture of student learning management: monograph. Kh. KhNURE, 235 (2004). (in Ukrainian)
5. Hlushman, T.M.: Components of the professional culture of a future organizational manager. Sci. J. "ScienceRise: Pedagogical Education" **5**(1), 8–13 (2016). (in Ukrainian)
6. Iashnyk, S.V.: Theoretical foundations of the formation of the management culture of future specialists in the agricultural industry. Bulletin of the Chernihiv National Pedagogical University named after T. G. Shevchenko. Series "Pedagogical Sciences". Vyp. 100. Chernihiv: ChNPU, pp. 451–454 (2012). (in Ukrainian)
7. Prystupa, V.V.: Theoretical and methodological support for the formation of economic competence of radio communication engineers. Bulletin of the National Aviation University 2, pp. 181–187 (2011). (in Ukrainian)
8. Soroka, O., Kalaur, S., Vaniuha, L.: Educational paradigm of training managers in the sphere of culture: nowadays challenges. Soc. Work. Educ. **8**(4), 515–527 (2021)
9. Kalaur, S., Soroka, O.: Theoretical and methodological principles of the formation of professional culture and professional image among educators by means of the acmeological approach in war conditions. Soc. Work. Educ. **10**(1), 99–110 (2023)
10. Brovko, K.A., Ternopilska, V.I.: Corporate culture of personality: psychological aspects. Topical issues of contemporary science: collection of scientific articles. C.E.I.M., Valencia, pp. 176–178 (2017)
11. Honchar, L.V.: Components of the culture of professional activity of future managers. Theory Methodol. Teach. (from fields of knowledge) **22**(1), 109–112 (2020). (in Ukrainian)
12. Mykhailiuk, N.V.: Professional culture of a specialist. Economic development and the legacy of Semyon Kuznets: materials of the International Scientific and Practical Conference. Kharkiv: KhNEU, pp. 366–367 (2019). (in Ukrainian)

13. Bilous, O., Hovorun, T., Berladir, K., Dunaeva, M.: Ensuring the quality of training engineers in a virtual environment. In: Tonkonogyi, V. (eds.) Advanced Manufacturing Processes II. InterPartner 2020. Lecture Notes in Mechanical Engineering, pp. 765–774. Springer, Cham (2021). https://doi.org/10.1007/978-3-030-68014-5_74
14. Mamchych, T., Myroniuk, L., Royko, L.: Information and communication technologies usage experience at teaching of mathematical disciplines in the terms of the distance learning. Comput.-Integr. Technol. Educ. Sci. Prod. **39**, 70–77 (2020). https://doi.org/10.36910/6775-2524-0560-2020-39-12
15. Hernandez-de-Menendez, M., Morales-Menendez, R.: Technological innovations and practices in engineering education: a review. Int. J. Interact. Des. Manuf. **13**, 713–728 (2019). https://doi.org/10.1007/s12008-019-00550-1
16. Titova, O., Ishchenko, T., Yershova, L., Bazyl, L., Kruchek, V.: Entrepreneurial mindset development in engineering students through a business canvas approach. In: Ivanov, V., Trojanowska, J., Pavlenko, I., Rauch, E., Piteľ, J. (eds.) Advances in Design, Simulation and Manufacturing VI. DSMIE 2023. Lecture Notes in Mechanical Engineering, pp. 400–409. Springer, Cham (2023). https://doi.org/10.1007/978-3-031-32767-4_38
17. Symonenko, S., Zaitseva, N., Titova, O., Vynogradova, M.: Development of communicative competence as a precondition of competitive software engineer formation. In: Nadykto, V. (eds.). Modern Development Paths of Agricultural Production, pp. 307–315. Springer, Cham (2019). https://doi.org/10.1007/978-3-030-14918-5_32

Author Index

F. J. G. Silva et al. (Eds.): DSMIE 2025, LNME, pp. 171–172, 2025.
https://doi.org/10.1007/978-3-031-96413-8

The manufacturer's authorised representative in the EU is Springer Nature Customer Service Centre GmbH, Europaplatz 3, 69115 Heidelberg, Germany. If you have any concerns regarding our products, please contact ProductSafety@springernature.com

Printed and bound by CPI Group (UK) Ltd, Croydon, CR0 4YY
01/10/2026
02225911-0020